WHAT DO THE WEEDS TELL ABOUT THE SOIL?

OBSERVE 191 WILD PLANTS,
DECODE SOIL SCIENCE,
IMPROVE YOUR GARDENING

BY MARIE CACHET

Dedicated to Catarina

X: @CachetMarie
YT: Marie Cachet

ISBN: 978-2-9592836-1-1
Dépot Légal Septembre 2024

INTRODUCTION

Wild plants, often perceived as weeds, are in fact valuable allies for gardeners and farmers. They possess a secret language, a natural code that, when deciphered, can reveal crucial information about the characteristics and quality of the soil. This book aims to introduce you to this science by showing you how to "read" these plants to understand the soil on which they grow.

Weeds are natural bio-indicators. Their presence, vigor, and diversity can inform us about specific soil conditions, such as its texture, composition, pH, and nutrient richness. By observing these plants, we can obtain a clear picture of the soil's state without the need to disturb it with traditional analysis methods, which can be invasive and costly.

The Objectives of This Book:

- **Discover the Role of Bio-Indicator Plants:**

Learn how spontaneous wild plants can give us valuable clues about the conditions and history of the soil, without even opening it. You will find pages explaining the bio-indicator characteristics of more than 190 of the most common plants in the northern hemisphere (particularly in Europe and North America). These pages are arranged alphabetically by the Latin name of the plants to avoid any confusion. The common name is indicated in parentheses, and at the end of this book, you will find both a glossary based on the Latin names and on the English or common names. For convenience, I have assigned each letter a color, making it easy to search for plants by looking at the edge of the book.

The soil functions as a hidden, living ecosystem, and I invite you to see the plants on its surface as a witness to this inner functioning. The more common a plant is in a given environment, the more relevant its bio-indicator characteristics will be for analysis.

- **Read Nature and Improve Gardening and Farming Practices:**

The purpose of this book is also to use the acquired knowledge to optimize soil management and gardening practices through this reading of wild plants. This knowledge will allow you to read Nature around you, analyze land, find water sources or even archaeological sites, or recognize polluted areas.

We will explore how to observe and interpret the signs that wild plants give us. By familiarizing yourself with the characteristics of more than 190 common plants in Europe and North America, you will be able to decode the messages these plants convey about the soil's condition.

The following introduction aims to give you the basic keys to understanding the terms used, the functioning of the soil, and the conditions for seed dormancy break of the plants detailed later on.

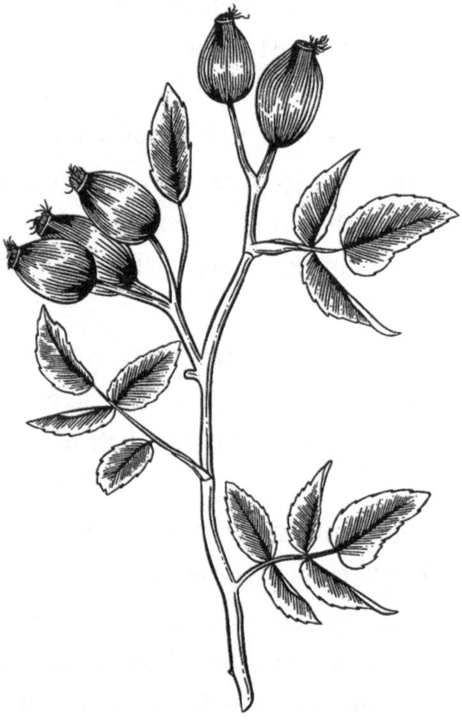

I. HOW DOES A SOIL WORK?

The soil is a living, dynamic, and complex ecosystem where numerous organisms work together to break down organic matter and recycle nutrients. This "digestion" of the soil is essential for maintaining its fertility and health.

1. Role of Animals, Fungi, and Mycorrhizae

Soil Animals: Earthworms, insects, and other small animals fragment and pre-digest organic matter and improve the soil's structure through their movements. Their activity creates aeration channels and promotes water drainage. They are often well-known to gardeners, as they can be easily observed. However, they are not always appreciated, such as slugs, which nonetheless play a valuable role in the construction and maintenance of healthy soil.

Fungi and Mycorrhizae: Fungi break down complex organic materials into simpler elements. Mycorrhizae, the network of fungal filaments in the soil, are the site of symbiotic associations between fungi and plant roots, increasing the surface area of root absorption and improving access to nutrients. Thus, in healthy soil, if an animal dies and decomposes in a specific location, it will actually nourish not just the small area where it died, but a vast zone, as nutrients are distributed by the mycorrhizae. In the absence of these mycorrhizae, the accumulation of organic matter in one place can quickly overwhelm the soil's digestion capacity and create pollution.

Mycorrhizae also act like sponges, contributing to the soil's water retention and distribution. They also serve as a communication pathway for plants, as a vegetal "internet", through their white filaments.

2. Aerobic and Anaerobic Bacteria

Aerobic Bacteria: These bacteria require oxygen to survive, so they live in the upper part of the soil and play a crucial role in the decomposition of organic matter and the release of essential nutrients like nitrogen. Aerobic bacteria contribute to the formation of soil aggregates. They produce sticky substances, such as polysaccharides, that bind soil particles together.

A good soil structure, with well-formed aggregates, creates pores of various sizes. These pores retain water and air, thereby improving soil aeration and water retention capacity.
In well-aerated and not compacted soils, aerobic bacteria consume oxygen and produce carbon dioxide, helping to maintain a balance between air and water in the soil. This process prevents the soil from becoming too waterlogged, which could otherwise promote anaerobic conditions.

Anaerobic Bacteria: These bacteria thrive in the absence of oxygen and are often responsible for decomposition in waterlogged environments, where they cause processes like putrefaction or peat formation. They can produce compounds like methane during the breakdown of organic matter and reduce nitrates to gaseous nitrogen. Anaerobic bacteria are also naturally present in the deeper layers of the soil, where they help bring up minerals from the bedrock beneath the earth. Some anaerobic bacteria produce indeed organic acids and other metabolites that can dissolve minerals from the bedrock, thereby releasing nutrients such as iron, manganese, calcium, magnesium, and phosphorus into the soil.
Anaerobic bacteria, like sulfate-reducing bacteria, play also a role in reducing sulfates to sulfides. This process can release minerals containing sulfur, making this essential element available to plants. The biological weathering of bedrock, where bacteria play a key role, is essential for soil development and the release of nutrients that are initially locked within minerals. By altering the soil chemistry in oxygen-poor environments, anaerobic bacteria contribute to the gradual transformation and release of these elements.

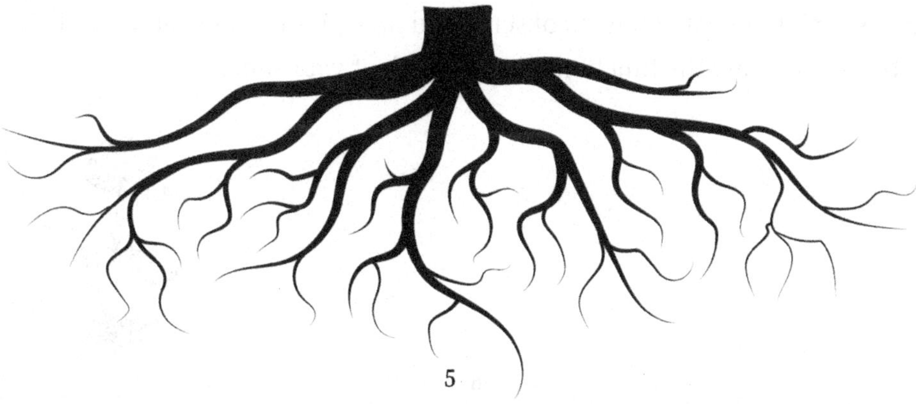

Understanding this highlights the importance of avoiding the mixing of soil layers over large areas, as in deep plowing. Such practices disturb or even kill large numbers of soil animals, as evidenced by the crows that come to feed themselves after plowing. The mycorrhizae and the fungal network they form are destroyed, and the soil layers are heavily mixed, bringing aerobic bacteria into anaerobic environments where they cannot survive, and anaerobic bacteria into aerobic environments where they cannot survive either.
Moreover, this doesn't account for the soil compaction caused by heavy machinery and tractors, which harms soil life, or the erosion triggered by disrupting the soil's natural balance and exposing it to the elements.
It is crucial not to overly disturb the natural layers of the soil. Each layer has a specific function, and intense mixing can harm the living organisms and biodiversity in the environment, degrade soil fertility, and even threaten the soil's stability. However, small disturbances are not problematic, as they mimic natural processes and can be quickly repaired if the surrounding soil is healthy.

3. How to Understand the Soil Without Opening It?

The key to this book is to show you how to deduce the characteristics of the soil by observing the plants that grow on it. Wild plants, often perceived as weeds, are in fact natural indicators of soil conditions. They serve as witnesses to the state of the soil. Each plant has specific preferences regarding soil type, pH, nutrient content, and moisture levels. By learning to recognize these plants and understand their needs, as well as the conditions required for their seeds to break dormancy, you can gain valuable insights into your soil.
Let us explore in detail how to observe and interpret bio-indicator plants to better understand the functioning and health of your soil...

II. FUNCTIONING OF A PLANT

1. Leaves and Buds

Leaves and buds are essential for the survival and growth of plants. They play crucial roles in photosynthesis and respiration, two vital processes for plants.

Respiration: Plants absorb carbon dioxide (CO_2) and release oxygen (O_2) through their leaves. This process is reversed during the night. Respiration allows plants to convert the sugars produced by photosynthesis into usable energy for growth and development.

Photosynthesis: Leaves function like solar panels, capturing sunlight to produce sugars from CO_2 and water (H_2O) instead of electricity. The sap transports these sugars to different parts of the plant, providing the raw material for growth.

Plant Matter: The majority of a plant's mass comes from the air, primarily from CO_2, rather than from the soil.

2. Roots

Roots anchor the plant in the soil and are essential for the absorption of water and nutrients.

Anchorage in the Soil: Roots stabilize the plant, helping it withstand adverse weather conditions.

Water Absorption: Roots absorb water from the soil, which is transported to the leaves to prevent drying out in the Sun and facilitate photosynthesis.

Absorption of Minerals and Nitrogen: Roots also absorb minerals and nitrogen necessary for growth. Nitrogen is particularly crucial for the production of proteins and chlorophyll (the green pigment that absorbs light during photosynthesis).

Root Exudates and Exchanges: Roots secrete substances called exudates, which nourish bacteria and mycorrhizae in the soil. These interactions promote the exchange of nutrients between the soil and the roots, and can either inhibit or encourage the growth of other plants.

3. Carbonaceous Trunk or Stem

The trunk or stem serves as a structural support and transports nutrients and water throughout the plant.

Skeleton and Support: The stem or trunk provides the necessary structure to keep the plant upright, allowing it to grow tall to get enough light be pollinated or to disperse its seeds.

Protection: The trunk or stem protects the plant's internal parts from mechanical damage and diseases.

Sap Transport: Raw sap, consisting of water and minerals, is transported from the roots to the leaves, while elaborated sap, rich in sugars, circulates from the leaves to the rest of the plant and the organs that use these sugars.

4. Flowers, Fruits, and Seeds

Flowers, fruits, and seeds are essential for the reproduction of plants.

Flowers contain the reproductive organs and are pollinated by wind, insects, or other means.

Fruits and Seeds: After pollination, flowers transform into fruits, which contain seeds. The seeds are dispersed to colonize new areas.

Seeds break their dormancy under specific conditions. Germination provides valuable information about the soil, as the soil often contains dormant seeds ready to germinate when conditions are favorable. Once germinated, plants may die, survive, or thrive depending on the surrounding conditions. Dominant plants, therefore, give us important insights into soil moisture and many other environmental factors.

III. CONDITIONS FOR BREAKING DORMANCY AND COLONIZING AN ENVIRONMENT

The conditions required for a plant to break dormancy and colonize an environment provide valuable insights into the surrounding environment and soil without the need to physically examine it. Understanding these conditions allows for a better comprehension of the soil and the surrounding ecosystem, and can even possibly be used to effectively fight undesirable plants by altering the conditions that favor their growth, rather than attempting to eradicate them through other means.

1. Climate

Some seeds will never germinate in an unsuitable climate, even if they are conserved in the soil for a long time. For example, a mango seed will never alone grow into a mango tree in Europe or North America, no matter how long it remains in the soil because the climatic conditions are not favorable. While climate is a crucial factor for seed germination, it is generally better understood through other means than the observation of wild plants. However, in some cases, the presence of some kinds of plants can provide clues about local microclimates.

2. Shaded or Sunny Environment

The availability of light is essential for plant growth, and bio-indicator plants can reveal areas that are more or less exposed to sunlight.

Plants that prefer shaded or sunny environments can help us understand the variations in light across a piece of land throughout the year. This can be particularly useful for assessing a terrain that is not well-known beforehand. By observing which plants dominate in certain areas, one can infer whether those spots receive more shade or sunlight, providing valuable information for planning and managing the land effectively.

3. Water Availability

The availability of water in an environment is sometimes obvious, but not always. Studying the wild plants present on a site allows us to understand the water availability throughout the year, the presence or absence of drought periods, and the hydric contrasts.

Wild plants can reveal the presence of underground water, the emergence of springs, as well as the quality and cleanliness of the water. They can also indicate the soil's capacity to retain water. By observing the types of plants thriving in an area, one can deduce whether the soil retains moisture well or tends to dry out fast, providing insights into the water dynamics of the landscape.

4. Soil pH

Soil pH measures the acidity or alkalinity of the soil, determined by the concentration of hydrogen ions (H^+) in the soil solution.

The pH scale ranges from 0 to 14. A pH of 6.5 to 7 is neutral; a pH below 6.5 to 7 is acidic, and a pH above 6.5 to 7 is alkaline. Soil pH is crucial because it influences the availability of nutrients to plants.

- Low pH indicates a high concentration of hydrogen ions (H^+), making the soil acidic.
- High pH indicates a low concentration of hydrogen ions (H^+), making the soil alkaline.

Factors Influencing Soil pH:

- Temperature: Temperature variations influence chemical reactions in the soil, thus affecting the pH. Cold temperatures tend to lower the pH.
- Moisture: Rainwater, often slightly acidic, can lower the soil pH. Evaporation of water can concentrate basic salts, increasing the pH.
- Time of Day and Seasons: The pH can fluctuate throughout the day and across seasons due to changes in biological and chemical processes. The pH tends to drop towards the end of the day and during winter.

- Biological Activities: Plant roots and soil microorganisms excrete substances that can alter the pH.
- Soil Amendments: The addition of organic matter, lime (to raise pH), or sulfur (to lower pH) can influence the soil pH.

Consequences of pH Changes:

- Low pH (Acidic Soil): Can limit the availability of essential nutrients like calcium, magnesium, and potassium but may improve the availability of other nutrients. Very acidic soils (with a pH below 4.5) can also release toxic aluminum ions harmful to plants.
- High pH (Alkaline Soil): Can make some nutrients, such as iron, manganese, zinc, and copper, less available to plants, potentially leading to deficiencies.

5. Organic Matter

Organic matter is essential for soil health and plays a central role in soil fertility, water retention, and nutrient availability. It is primarily composed of plant-based organic matter (mainly rich in carbon) and animal-based organic matter (mainly rich in nitrogen, though it varies from one animal to another), each with specific characteristics and roles in the soil ecosystem.

Organic matter in the soil comes from the decomposition of plant residues (leaves, roots, wood) and animal residues (excrement, carcasses, insect remnants). It consists of fresh residues and decomposed organic matter (humus). Humification, or the transformation of organic matter into humus, is a key process for soil health.

Role of Organic Matter in the Soil:

- **Soil Structure:** Organic matter, especially from plant residues, improves soil structure by increasing its porosity, which allows for better infiltration of water and air.

- **Water Retention:** Soils rich in organic matter have a better capacity to retain water, particularly thanks to plant-based organic matter, which acts like a natural sponge. However, if present in excessive quantities and not balanced with nitrogen-rich matter, it may not decompose properly, leading to imbalances. For example, it can form a mat on the soil surface, preventing the circulation of water, air, and essential nutrients, and can acidify the soil.
- **Nutrient Availability:** The balanced decomposition and digestion of organic matter, whether plant-based or animal-based, release essential nutrients (nitrogen, phosphorus, sulfur, etc.) in a form that plants can absorb.
- **Microbial Activity:** Organic matter feeds soil microorganisms, which play a central role in the decomposition of organic matter and the nutrient cycle. Animal excrement, rich in nitrogen, particularly stimulates microbial activity.

Organic matter is crucial for creating a healthy, fertile soil environment that supports plant growth and sustains the broader ecosystem.

Nitrogen (N) in Organic Matter:

- Plant Origin: Nitrogen in plant-based organic matter primarily comes from the proteins in green plant residues. Decomposing green leaves and stems gradually release this nitrogen into the soil.
- Animal Origin: Animal excrement is a rich and concentrated source of nitrogen. Poultry manure, for example, is very high in nitrogen, followed by cow, sheep, and horse manure. Nitrogen from animal organic matter is generally more readily available to plants than that from plant residues.

Carbon (C) in Organic Matter:

- Plant Origin: Carbon is primarily stored in plant residues, especially in the brown parts (straw, wood), in the form of cellulose, hemicellulose, and lignin. These compounds are more resistant to decomposition, allowing carbon to remain in the soil for a long time as humus.
- Animal Origin: Carbon in animal organic matter is also present, mainly in the form of lipids and proteins. However, its decomposition is often faster than that of plant carbon. Horse manure is a type of excrement that is particularly rich in carbon.

Differences Between Animal Manures:

- Poultry Manure: Very rich in nitrogen, with a low C/N ratio, meaning it decomposes quickly and releases nutrients immediately.
- Cow Manure: Contains a good balance of nitrogen and carbon, with a moderate C/N ratio. It decomposes steadily, gradually enriching the soil.
- Horse Manure: Rich in fibers (carbon), with a higher C/N ratio, it decomposes more slowly, providing a long-lasting effect on soil structure.
- Sheep Manure: Nutrient-dense and decomposes quickly, though in a more moderate amount compared to poultry manure.
- Goat Manure: Relatively rich in nutrients with a good balance between nitrogen and carbon. It decomposes quickly and improves soil structure while being less compact and easier to handle than some other manures.
- Rabbit Manure: Very rich in nitrogen, phosphorus, and potassium, it decomposes quickly. Its low C/N ratio makes it an excellent amendment for rapidly supplying nutrients, especially in poor soils. However, it is richer in carbon than poultry manure.
- Duck and Goose Manure: These manures are also rich in nitrogen, with a low C/N ratio, similar to that of other poultry. They decompose quickly and are particularly beneficial for crops that require a high nitrogen input.

Nitrogen (N) and Carbon (C):

The cycles of nitrogen and carbon are fundamental to the functioning of terrestrial ecosystems and plant growth.

Nitrogen (N):

- Role in Plants: Nitrogen is crucial for plant growth. It is a fundamental component of amino acids, which form proteins, and nucleic acids (DNA and RNA), essential for cell growth. Nitrogen is also a constituent of chlorophyll, the molecule responsible for photosynthesis.
- Available Forms of Nitrogen: Plants absorb nitrogen primarily in the form of nitrate (NO_3^-) and ammonium (NH_4^+). These forms are produced by the decomposition of organic matter, biological nitrogen fixation by certain bacteria, and the application of fertilizers.
- Nitrogen Cycle: The nitrogen cycle includes several stages: fixation (by legumes and symbiotic bacteria), mineralization (decomposition of organic matter), nitrification (conversion of ammonium to nitrate), and denitrification (return of nitrogen to the atmosphere as N_2).
- Indications by Plants: Plants such as clover and legumes indicate nitrogen-rich soil, while plants like horsetail can signal a lack of available nitrogen.

Carbon (C):

- Role in Plants: Carbon is the cornerstone of life on Earth. It is the main component of carbohydrates, which serve as an energy source for plants and other organisms. Carbon is also present in lipids, proteins, and nucleic acids.
- Photosynthesis and Carbon Fixation: Through photosynthesis, plants absorb CO_2 from the air and use solar energy to convert it into carbohydrates. This process is the foundation of plant biomass production.

Carbon Cycle: Carbon circulates between the atmosphere, soil, and living organisms through photosynthesis, respiration, decomposition, and combustion. Organic matter in the soil, particularly humus, represents a significant carbon reservoir.

Interactions Between Nitrogen and Carbon:
C/N Ratio: The carbon-to-nitrogen ratio (C/N) is an important indicator of organic matter decomposition. A high C/N ratio means that organic matter decomposes slowly, which can limit the availability of nitrogen for plants. A balanced C/N ratio promotes efficient decomposition, releasing both carbon and nitrogen into the soil.

Understanding the balance and interaction between nitrogen and carbon in the soil is key to managing soil fertility and supporting healthy plant growth. The richness and balance of organic matter can be assessed by the presence of some specific plants.

6. Mineral Salts, Bases, and Pollutions

Minerals, although they constitute a small proportion of a plant's mass, are essential for its growth and health.

Bases

Soil bases are basic ions present in the soil, mainly calcium (Ca^{2+}), magnesium (Mg^{2+}), potassium (K^{+}), and sodium (Na^{+}). They are often associated with the soil's ability to neutralize acidity and maintain a balanced pH, and they are themselves essential nutrients for plants. The soil bases (calcium, magnesium, potassium, and sodium) primarily come from the breakdown of rocks and minerals found in the soil. This process releases mineral elements that then become available to plants. They can also come from fertilizers and amendments applied by humans, as well as from decomposing organic matter that releases these nutrients as it breaks down. An acidic soil may contain these bases, but they are often less available to plants. In an acidic soil, the high concentration of protons (H^{+}) and sometimes aluminum (Al^{3+}) can interfere with the absorption of bases by plant roots, limiting their availability. Rainfall and infiltrating water can also lead to a loss of bases through leaching, as H^{+} ions (protons) promote the release of basic ions, which are then carried away by the water.

Role of Minerals in Plants:

Potassium (K): Regulates the opening and closing of stomata, thus helping to control transpiration and the plant's water balance. It is also crucial for protein synthesis and photosynthesis.

Phosphorus (P): Essential for the formation of DNA, RNA, and cell membranes. It plays a key role in energy storage and transfer (ATP).

Zinc (Zn): Important for protein synthesis, growth, and development of plants. It is a crucial component of many enzymes.

Calcium (Ca): Necessary for cell wall growth and cell division. It also aids in cellular signaling.

Magnesium (Mg): A central component of chlorophyll, indispensable for photosynthesis.

7. Other Aspects

The breaking of seed dormancy and the presence of certain plants can also provide insights into the ecological history of a site and its current biodiversity.

- Archaeophyte Plants: These plants, which are species that were most often introduced to an area in ancient times and have since naturalized, can reveal historical information about a site. Their presence may indicate past human activity or long-established ecosystems.
- State of Biodiversity: The diversity of bio-indicator plants can offer clues about the health and resilience of the ecosystem. A rich variety of plants suggests a healthy, balanced ecosystem, while a lack of diversity might indicate environmental stress or degradation.

These additional aspects help to deepen the understanding of a site's ecological context, offering a broader perspective on the history of the soil and the current environmental status.

ACHILLEA MILLEFOLIUM (YARROW)

SOIL EROSION.
DISTURBED OR DEGRADED SOIL.
TOLERANCE TO DROUGHT & FULL SUNLIGHT
MODERATE SALINITY.
PH OF 5.0 TO 8.0
MINOR CONTAMINATION WITH HEAVY METALS

Indications of Soil Erosion:

It prefers well-drained or excessively-drained soil. It usually grows where soils are expected to leach out and erode, often due to agriculture, either from plowing or overgrazing.

Yarrow is drought-tolerant, making it a good indicator of habitats where periods of drought or low availability of water are normally expected.

Adapted to a Wide Range of Soils:

Yarrow grows in loamy, sandy, and clay soils that are poor in nutrients but rich in bases.

Salinity Tolerance:

It tolerates moderate salinity in the soil; hence, its presence can indicate either natural soil salinity or soils that have been watered under hot weather, the consequence of that being salt from the water appearing in the soil through evaporation.

Soil pH:

Yarrow grows best in soils that are slightly acid to alkaline (5.0 to 8.0 pH).

Exposure to Sunlight:

Yarrow is a sun-loving plant, and is often found in sunny habitats like meadows, roadside edges, and fields. the presence of Yarrow shows that full light is available.

Disturbance Indicator:

This species is commonly found in disturbed localities such as abandoned farmlands, fallow lands, open areas of crops, and vacant lots. It often indicates a history of recent or historical habitat disturbance.

Tolerance to Heavy Metals:

This plant is able to tolerate some amount of heavy metals.

It can serve as an indicator of soils contaminated with such substances when associated with plants of an equivalent nature.

AEGOPODIUM PODAGRARIA (GROUND ELDER)

RICH SOIL
WATERLOGGED SOIL
SHADE
PH OF 6.0 TO 7.5

Preference for Forest Soils:
Ground elder does well in soils rich in organic matter, nutrients and bases. It indicates a fertile soil.

Indication of Waterlogged Soils:
The plant indicates good availability of water; it even suggests the soil tends to be waterlogged.

Shade Tolerance:
Ground elder can grow under partial to full shade conditions. Its primary habitat is, indeed, in the gloomy light of the forest edge. It shows low direct sunlight exposure.

Colonization Ability:
Aegopodium podagraria has strong abilities to colonize and quickly spread through its rhizome. This is, indeed, its main reproduction way, like for almost all the herbs growing in the forest. If the surrounding meets its requirements, it can easily become dominant.

Preference for Neutral to Slightly Alkaline Soils:
It is well adapted to soils with neutral–slightly alkaline pH (pH 6.0 to 7.5). This plant might thus be a sign of a non-acidic soil pH.

AJUGA REPTANS (BUGLE)

MODERATELY MOIST TO MARSHY SOILS
ORGANIC MATTER- AND NUTRIENT-RICH SOILS
PARTIALLY SHADOWED TO FULL SHADE CONDITIONS
PH OF 6.0 TO 7.5
DISTURBED HABITAT

Preference to Moist Soils:

Ajuga reptans usually does well in moist and clayey soil. This marsh and bog plant is indicative of water-logged conditions and hydromorphisms (lack of oxygen because of the water-logged nature of the soil) especially if growing with other equivalent plants. It can suggest soil work during wet weather or overgrazing in wet meadows. The soil becomes compacted .

Organic Matter and Nutrient-Rich Soil:

It prefers organic-rich soils, especially animal organic matter. Its presence can indicate an accumulation of organic matter that soil bacteria can no longer decompose because of the watery soil.

Shade Tolerance:

Bugleweed grows well under partial to full shade conditions. Its presence can indicate low direct sunlight exposure.

Tolerance to a Range of Soil pH:

It thrives in soils with a pH ranging from slightly acidic to neutral (pH 6.0 to 7.5). Its presence can indicate a stable and suitable soil pH, ideal for many plants.

Propagation by Stolons:

Ajuga reptans spreads rapidly by stolons, making it an effective colonizer of disturbed habitats.

ALBIES ALBA (SILVER FIR)

FRESH AIR QUALITY AND LOW POLLUTION
COLD AND WET CLIMATIC CONDITIONS
PH OF 4.5 TO 7
DEEP, WELL-DRAINED, AND NUTRIENT-RICH SOIL
REGULAR WATER AVAILABILITY

Sensitivity to Air Pollution:

White fir is sensitive to some air pollutants, notably sulfur dioxide and ozone. If one sees an abundance of white fir, it generally indicates that air quality is relatively good.

High Moisture Requirements:

This tree prefers deep soil with good drainage, always being moist, and high air humidity.

Tolerance to Cold:

Abies alba is cold-tolerant; hence, it is often distributed in cold zones or in mountains, usually within 300 to 2000 m, with its presence normally being an indication of pretty tough winters.

Deep Soil:

Abies alba grows best on deep, nutrient-rich, well-drained soils.

Soil Preference:

It rather prefers soils that are slightly acidic to near neutral, with a pH ranging from about 4.5 to 7.

Drought Sensitivity:

This specific tree does not do well in drought conditions, and its presence normally indicates a constant supply of water even during the summer. This shows places with consistent rainfalls or good water retention in the soils.

ALNUS GLUTINOSA (BLACK ALDER)

MOIST OR WATERLOGGED SOILS
NITROGEN FIXATION
TOLERANCE TO VARIATIONS IN SOIL PH
PH USUALLY OF 4.5 TO 7.5
PERIODIC FLOODING
DISTURBED AND UNSTABLE SOILS

Preference for Moist to Waterlogged Soils:

Black alder prefers moist soils and will often be found by watercourses, in swamps, and in flood-prone areas. Black alder is indicative of heavy and water-saturated soils, most often being hydromorphic.

Nitrogen Fixation:

Black alder helps in nitrogen fixation with the help of symbiotic root nodules through free-living soil bacteria (Frankia alni). This nitrogen is made available to other plants growing in its vicinity.

Tolerance to a Wide Range of Soil pH:

Sustained growth has been noticed under acidic to slightly alkaline soils (pH 4.5 to 7.5). This tree is very tolerant to pH variation, though it prefers neutral to slightly acidic soils.

Flood Resistance:

It tolerates flooding and waterlogged soils well. Its presence often indicates areas that experience periodic flooding or are frequently waterlogged. Its roots tolerate these floods for long periods without rotting.

Indication of Disturbance:

The black alder is a pioneer tree, often found in disturbed habitats, such as riverbanks or areas where the soil has been churned. Its presence can indicate natural regeneration after a disturbance.

Unstable Soil:

In many cases, it is an indicator of unstable soil. It helps to stabilize riverbanks, thus preventing erosion, and can lead to improved soil stability and reduced erosion.

ALTHAEA OFFICINALIS (MARSHMALLOW)

MOIST TO WATERLOGGED SOILS
RICH SOILS
SALT RISE
PH OF 6.5 TO 8.0
OPEN OR SEMI-OPEN AREAS
LIGHTLY DISTURBED HABITATS

Preference for Moist Soils:

Marshmallow thrives in moist to waterlogged soils.

Tolerance for Soils Rich in Organic Matter:

This plant prefers soils rich in organic plant matter and bases. It is very resistant to salt excesses and often indicates a salt rise, either near coastal areas or due to excessive irrigation in hot weather.

Preference for Neutral to Slightly Alkaline Soils:

Althaea officinalis prefers soils with a pH ranging from neutral to slightly alkaline, i.e., a pH range of 6.5 to 8.0. Its presence can indicate soil with a stable and often slightly basic pH.

Preference for Sunny to Partially Shaded Environments:

Marshmallow thrives in sunny to partially shaded environments. Its presence can indicate open or semi-open areas with good light exposure.

Tolerance to Light Disturbances:

Althaea officinalis can tolerate light disturbances and colonize lightly disturbed areas. Its presence may indicate a habitat that has undergone moderate natural or human disturbances.

AMARANTHUS BLITUM (GUERNSEY PIGWEED)

WELL-DRAINED TO MODERATELY MOIST SOILS
HIGHLY FERTILE SOILS AT THE TIME OF GERMINATION
PH OF 5.5 TO 7.5

OPEN AREAS

PERIODS OF DROUGHT
SOIL SALINIZATION, OFTEN DUE TO IRRIGATION
DISTURBED HABITATS

Preference for Well-drained to Moderately Moist Soils:
Amaranthus blitum thrives in well-drained to moderately moist soils. Its presence often indicates soil that does not retain excessive water, allowing good root aeration.

Tolerance to Soils Rich in Organic Matter and Nutrients:
This plant like other amaranths prefers soils very rich in organic matter and nutrients. It grows well after soil has been enriched with various fertilizers: potassium, phosphorus, or nitrogen.
The high presence of these elements in the surface soil, or the high presence of one of them, triggers the germination of its dormant seeds. Fertilizer application after the beginning of its growth does not significantly affect its development, but it is crucial for its germination.
Fertilizers can be artificial or natural, such as after the passage of birds or poultry, whose droppings are very rich in nitrogen.
The germination of this amaranth highlights soil salinization, often caused by irrigation in hot weather, which concentrates the mineral salts of the water on the soil surface during evaporation.
It should also be noted that amaranths are resistant to most pesticides, and their presence may indicate agricultural soil treated with pesticides.
Its presence can thus indicate excessively fertilized soil, often found in gardens, vegetable patches, cultivated fields, animal parks, roadsides, and fallow lands.

Preference for Slightly Acidic to Basic Soils:
Guernsey pigweed prefers soils with a pH ranging from slightly acidic to basic, i.e., a pH range of 5.5 to 7.5.

Preference for Sunny Environments:
Amaranthus blitum thrives in sunny environments. Its presence can indicate open areas with great light exposure.

Tolerance to Disturbances:
Amaranthus blitum is relatively tolerant to disturbance and can quickly colonize disturbed areas. Its presence indicates a habitat that has recently undergone natural or human disturbances.

AMARANTHUS HYBRIDUS (GREEN AMARANTH)

AND OTHER AMARANTHS

WELL-DRAINED TO MODERATELY MOIST SOILS
HIGHLY FERTILE SOILS AT THE TIME OF GERMINATION
PH OF 5.0 TO 7.5
OPEN OR SEMI-OPEN AREAS
PERIODS OF DROUGHT
DISTURBED HABITATS

Preference for Well-drained to Moderately Moist Soils:
Green amaranth thrives in well-drained to moderately moist soils. Its presence often indicates soil that does not retain excessive water, allowing good root aeration. This plant is relatively drought-resistant and often indicates environments with limited water availability at certain times of the year.

Tolerance to Soils Rich in Organic Matter and Nutrients:
This plant prefers soils that are very rich in organic matter and nutrients. It grows well after soil has been enriched with various fertilizers: potassium, phosphorus, or nitrogen. The high presence of these elements in the surface soil, or the high presence of one of them, triggers the germination of its dormant seeds. Fertilizer application after the beginning of its growth does not significantly affect its development, but it is crucial for its germination. Fertilizers can be artificial or natural, such as after the passage of birds or poultry, whose droppings are very rich in nitrogen.

It should also be noted that amaranths are resistant to most pesticides, and their presence may indicate agricultural soil treated with pesticides.

Its presence can thus indicate highly fertile or excessively fertile soil, often found in gardens, vegetable patches, cultivated fields, animal parks, roadsides, and fallow lands.

Preference for Slightly Acidic to Basic Soils:
PH range of 5.0 to 7.5.

Preference for Sunny to Partially Shaded Environments:
Green amaranth thrives in sunny to partially shaded environments.

Tolerance to Disturbances:
Amaranthus hybridus is relatively tolerant to disturbance and can quickly colonize disturbed areas. Its presence indicates a habitat that has recently undergone natural or human disturbances.

ANEMONE NEMOROSA (WOOD ANEMONE)

MOIST BUT WELL-DRAINED AND AERATED SOILS
SOILS RICH IN ORGANIC MATTER
PH OF 5.0 TO 7.0
SHADED ENVIRONMENTS
STABLE ENVIRONMENT WITH LITTLE DISTURBANCE

Preference for Moist and Well-drained Soils:

Wood anemone thrives in moist but well-drained soils. Its presence often indicates a soil that retains moisture without being waterlogged, with good structure allowing proper water and air circulation, which ensures good root aeration.

Soils Rich in Organic Matter:

This plant prefers soils rich in organic matter, often found in woodlands where leaf litter decomposes into humus. Its presence can indicate soil rich in fossil organic matter (originating from sedimentary rocks) and bases.

Indication of Neutral to Slightly Acidic Soils:

Wood anemone thrives in soils with a pH that is neutral to slightly acidic (pH 5.0 to 7.0). Its presence indicates soil with a stable pH, suitable for many woodland plants.

Preference for Shaded Environments:

It mainly grows in shaded or partially shaded environments, such as deciduous forests and forest edges. Its presence is a sign of dense vegetation cover with low direct sun exposure. If not introduced, it typically inhabits old, stable, and undisturbed forests.

ANGELICA SYLVESTRIS (WILD ANGELICA)

WET TO WATERLOGGED SOILS
MARSHY AREAS
SOILS RICH IN ORGANIC MATTER AND NITRATES
ANAEROBIC SOIL CONDITIONS, HYDROMORPHISMS
PH OF 5.5 TO 7.0
PARTIALLY SHADED ENVIRONMENTS.

Preference for Wet to Waterlogged Soils:

Wild angelica thrives in wet and waterlogged soils. Its presence indicates good water availability in the soil, often found along rivers, in swamps, and wetlands. It suggests humid air and anaerobic soil conditions, with hydromorphisms, meaning soil pore saturation with water.

Tolerance to Soils Rich in Organic Matter:

This plant prefers soils rich in organic matter. Its presence can indicate an abundance of plant organic matter and nitrate.

Tolerance to Slightly Acidic to Neutral Soils:

Wild angelica grows well in soils with a slightly acidic to neutral pH (5.5 to 7.0).

Preference for Sunny to Partially Shaded Environments:

It grows in sunny to partially shaded environments, often found in wet meadows and forest edges. Its presence indicates an area with moderate light exposure.

Change in Soil Nature According to the Season:

Although it prefers very wet soils, it also requires drainage at certain times of the year to prevent root rot. Thus, its presence often indicates waterlogged soil in winter and better-drained soil in summer.

ANTHEMIS ARVENSIS (FIELD CHAMOMILE)

WELL-DRAINED SOILS
SOIL EROSION
ROCKY OR SANDY SOILS
POOR SOILS
PH OF 4.5 TO 6.5
OPEN OR SEMI-OPEN AREAS
DISTURBED HABITATS

Preference for Well-drained Soils:

Field chamomile thrives in well-drained soils. Its presence often indicates very thin, rocky, or sandy soil that does not retain water well, indicating soil erosion.

Tolerance to Poor to Moderately Nutrient-rich Soils:

This plant can develop in soils that are poor to moderately rich in nutrients. Its presence may indicate soil that is not particularly fertile, often found in cultivated fields, rocky areas, and sandy zones.

Preference for Acidic to Neutral Soils:

Anthemis arvensis prefers soils with a pH ranging from acidic to neutral, i.e., a pH range of 4.5 to 6.5. Its presence generally indicates a lack of lime in the soil.

Preference for Sunny to Partially Shaded Environments:

Field chamomile thrives in sunny to partially shaded environments in cool climates. Its presence can indicate open or semi-open areas with good light exposure.

Tolerance to Disturbances:

Anthemis arvensis is relatively tolerant to disturbance and can quickly colonize disturbed areas. Its presence often indicates a habitat that has undergone natural or human disturbances.

ANTHRISCUS SYLVESTRIS (COW PARSLEY)

MOIST BUT NOT WATERLOGGED SOILS
RICH SOILS
FRESH AND DEEP SOILS
PH OF 6.5 TO 8.0
PARTIALLY SHADED ENVIRONMENTS
DISTURBED HABITATS

Preference for Moist Soils:

Cow parsley thrives in moist soils. Its presence often indicates fresh soil that retains moisture well but does not remain waterlogged.

Tolerance to Soils Rich in Organic Matter and Nutrients:

This plant prefers deep soils rich in bases, organic plant matter, and relatively high in nitrogen. Its presence may indicate fertile soil, often found in meadows, roadsides, ditches, and forest edges.

Preference for Neutral to Slightly Alkaline Soils:

Anthriscus sylvestris prefers soils with a pH ranging from neutral to slightly alkaline, i.e., a pH range of 6.5 to 8.0. Its presence may indicate soil with a stable and slightly basic pH.

Preference for Sunny to Partially Shaded Environments:

Cow parsley thrives in sunny to partially shaded environments. It is often found in clearings, forest edges, and open meadows in cool climates.

Tolerance to Disturbances:

Anthriscus sylvestris is relatively tolerant to disturbance and can quickly colonize disturbed areas. Its presence may indicate a habitat that has undergone natural or human disturbances, such as abandoned farmland, fallow lands, and roadsides.

APIUM GRAVEOLENS (WILD CELERY)

WET TO WATERLOGGED SOILS
SOILS RICH IN ORGANIC MATTER
SOILS RICH IN SALTS
POSSIBLY A SIGN OF AGRICULTURAL CHEMICAL INPUTS
PH 6.0 TO 7.5
SUNNY ENVIRONMENTS
DISTURBED HABITATS.

Preference for Very Moist and Saline Soils:
Marsh celery primarily grows in wet to waterlogged soils, often found along rivers, in swamps, coastal marshes, and flood zones, particularly in saltwater. Its presence indicates high atmospheric humidity and the presence of saltwater.

Tolerance to Soils Rich in Organic Matter and Salts:
This plant grows in soils very rich in organic matter. It has a high tolerance to salt and can thus indicate salt rises in the groundwater table or soil salinization due to irrigation in hot weather or the application of agricultural chemicals.

Alkaline Soil pH:
Apium graveolens grows well in soils with a pH ranging from neutral to slightly alkaline (pH 6.0 to 7.5).

Preference for Sunny to Partially Shaded Environments:
It grows in sunny environments. Its presence thus indicates an area with good light exposure.

Indication of Moderate Disturbance:
This plant is often found in disturbed habitats, such as roadsides and cultivated areas. Its presence can indicate moderate habitat disturbance.

AQUILEGIA VULGARIS (COMMON COLUMBINE)

WELL-DRAINED SOILS
SOILS MODERATELY RICH IN NUTRIENTS
LACK OF ANIMAL ORGANIC MATTER
PH OF 6.0 TO OVER 7.5
PARTIALLY SHADED ENVIRONMENTS

Preference for Well-drained to Moderately Moist Soils:

The common columbine thrives in well-drained to moderately moist soils. Its presence often indicates soil that retains enough moisture but is never waterlogged.

Tolerance to Soils Moderately Rich in Nutrients:

This plant prefers soils moderately rich in available nutrients, but high in bases and fossil plant organic matter, often found in undergrowth and forest edges. Its presence may indicate a lack of animal organic matter and an evolution of the biotope towards the forest.

Preference for Neutral to Alkaline Soils:

Aquilegia vulgaris prefers soils with a neutral to basic pH, from 6.0 to above 7.5.

Preference for Partially Shaded Environments:

The common columbine grows well in partially shaded environments.

ARCTIUM LAPPA (GREAT BURDOCK)

WELL-DRAINED AND MODERATELY MOIST SOILS
NUTRIENT-RICH SOILS
SOIL PH OF 6.0 TO 7.5
SUNNY ENVIRONMENTS
DISTURBED HABITATS

Preference for Well-drained to Moderately Moist Soils:

Great burdock thrives in well-drained and moderately moist soils. Its presence often indicates soil that retains enough moisture without ever being waterlogged, and moderate hydric contrasts.

Soils Rich in Organic Matter:

This plant prefers nutrient-rich soils. Its presence indicates a soil where organic matter is well decomposed and diverse, often found in waste areas, fallow land, and compost. It grows well in soils rich in bases and woody material.

Preference for Neutral to Slightly Alkaline Soils:

Arctium lappa prefers soils with a pH ranging from neutral to slightly alkaline, typically between 6.0 and 7.5.

Preference for sunny to partially shaded environments:

Great burdock grows well in sunny to partially shaded environments. Its presence can indicate open or semi-open areas with good light exposure.

Tolerance to Disturbance:

Arctium lappa is highly tolerant to disturbance and can quickly colonize disturbed areas. Its presence often indicates habitats that have undergone natural or human disturbances, such as waste areas, construction sites, and roadsides.

ARMORACIA RUSTICANA (HORSERADISH)

MOIST TO WELL-DRAINED SOILS
SOILS RICH IN ORGANIC MATTER AND NUTRIENTS
PH OF 6.0 TO 8.0
SUNNY TO PARTIALLY SHADED ENVIRONMENTS
MODERATE DISTURBANCES

Preference for Moist to Well-drained Soils:

Horseradish thrives in moist to well-drained soils. Its presence often indicates soil that retains sufficient moisture without being waterlogged.

Soils Rich in Organic Matter:

This plant prefers soils rich in organic matter and nutrients. Its presence can indicate fertile soil, often found in gardens, cultivated lands, and along waterways.

Preference for Neutral to Slightly Alkaline Soils:

Armoracia rusticana prefers soils with a stable pH ranging from neutral to slightly alkaline, typically between 6.0 and 8.0.

Preference for Sunny to Partially Shaded Environments:

Horseradish grows well in sunny to partially shaded environments. Its presence can indicate open or semi-open areas with good light exposure.

Tolerance to Moderate Disturbances:

Armoracia rusticana is relatively tolerant of disturbances and can quickly colonize disturbed areas. Its presence may indicate habitats that have undergone natural or human disturbances.

ARTEMISIA ABSINTHIUM (WORMWOOD)

DRY SOILS
WIDE RANGE OF SOILS
OVERGRAZING
PH OF 7.0 TO 8.5
SUNNY ENVIRONMENTS
PERIODS OF DROUGHT
MODERATELY TO HIGHLY DISTURBED HABITATS

Preference for Dry to Moderately Moist Soils:

Wormwood thrives in dry to moderately moist soils. Its presence often indicates well-drained soil capable of tolerating dry conditions.

Tolerance to a Wide Range of Soils:

This plant can grow in nutrient-poor, rocky soils but thrives in nutrient-rich conditions, especially those rich in bases and animal organic matter (notably from sheep). Its presence can in this case indicate overgrazed and disturbed soils.

Preference for Alkaline to Neutral Soils:

Artemisia absinthium prefers alkaline to neutral soils (pH 7.0 to 8.5). Its presence can indicate higher pH soils, often calcareous or rich in bases.

Preference for Sunny Environments:

Wormwood grows well in environments with full sun exposure. Its presence indicates open areas with full light exposure.

Tolerance to Drought:

Artemisia absinthium is drought-resistant, making it a good indicator of environments that experience periods of drought or have low water availability.

Indication of Moderate to High Disturbance:

This plant is often found in disturbed habitats such as roadsides, wastelands, and fallow fields. Its presence can indicate habitat disturbance, often related to human activity or overgrazing.

ARTEMISIA VULGARIS (COMMON MUGWORT)

WELL-DRAINED SOILS
BASE-RICH SOILS
NITROGEN-RICH SOILS
COMPACTED SOILS
PH OF 6.0 TO 8.0
OPEN OR SEMI-OPEN AREAS
PERIODS OF LOW WATER AVAILABILITY.
SOIL EROSION.

Preference for Well-drained soils:

Common mugwort thrives in well-drained soils. Its presence often indicates soil that allows good aeration of the roots.

Tolerance to Moderately Rich Soils:

This plant can develop in soils poor in organic matter but very rich in bases and relatively rich in nutrients particularly in azote. Its presence indicates soil that has undergone some disturbance, especially compaction and that is prone to erosion. It generally grows where there is a lack of ground cover.

Preference for Neutral to Slightly Alkaline Soils:

Artemisia vulgaris prefers neutral to slightly alkaline soils (pH 6.0 to 8.0).

Preference for Sunny to Partially Shaded Environments:

It grows in environments well-exposed to sunlight or with partial shade. Its presence can indicate open or semi-open areas.

Tolerance to Drought:

Artemisia vulgaris is drought-resistant, making it a good indicator of environments that can undergo periods of drought or have low water availability.

Indication of Moderate to High Disturbance:

This plant is often found in disturbed habitats such as roadsides, wastelands, and fallow fields. Its presence can indicate significant habitat disturbance, often linked to human activity.

ASARUM EUROPAEUM (EUROPEAN WILD GINGER)

MOIST, WELL-DRAINED SOILS
BASE-RICH SOILS
PH OF 6.0 TO 7.5
SHADED ENVIRONMENTS.

Preference for Moist to Well-drained Soils:

European wild ginger thrives in moist to well-drained soils. Its presence often indicates soil that retains sufficient moisture without being constantly waterlogged.

Tolerance to Soils rich in Organic Matter:

This plant prefers soils rich in bases but poor in animal organic matter. Its presence can indicate the presence of iron, manganese and/or limestone.

Preference for Neutral to Slightly Alkaline Soils:

Asarum europaeum prefers soils with a pH ranging from neutral to slightly alkaline, typically between 6 and 7.5. Its presence may indicate soil with a stable pH, suitable for many forest plants.

Preference for Shaded to Partially Shaded Environments:

European wild ginger grows well in shaded to partially shaded environments. Its presence can indicate areas with good vegetation cover.

Tolerance to Light Disturbance:

Asarum europaeum can tolerate light disturbance and colonize slightly disturbed areas. Its presence may indicate a habitat that has undergone moderate disturbances.

ASCLEPIAS SYRIACA (COMMON MILKWEED)

WELL-DRAINED SOILS
SANDY POOR SOILS
BASE-RICH SOILS
PH ABOVE 7.5
SUNNY ENVIRONMENTS
DISTURBED HABITAT.

Preference for Well-drained to Dry Soils:
Common milkweed thrives in well-drained to dry soils. Its presence often indicates sandy soil that doesn't retain moisture. It can indicate periods of drought.

Tolerance to a Wide Variety of Soils:
This plant can develop in soils poor in organic matter but moderately rich in nutrients and bases. Its presence can indicate soil that is not particularly fertile, often found in meadows, roadsides, and vacant lots.

Preference for Alkaline Soils:
Asclepias syriaca prefers soils with an alkaline pH, above 7.5.

Preference for Sunny Environments:
Common milkweed grows well in environments with full sun exposure.

Tolerance to Disturbance:
Asclepias syriaca is highly tolerant of disturbance and can quickly colonize disturbed areas.
Its presence indicates a habitat that has undergone natural or human disturbances.

BELLIS PERENNIS (DAISY)

WELL-DRAINED TO MODERATELY MOIST SOILS
NUTRIENT-POOR TO MODERATELY NUTRIENT-RICH SOILS
SOILS SUBJECT TO EROSION
SOIL DECALCIFICATION
PH OF 6.0 TO 7.5
SUNNY TO PARTIALLY SHADED ENVIRONMENTS
TRAMPLED SOIL

Preference for Well-drained to Moderately Moist Soils:
Daisies thrive in well-drained to moderately moist soils. Their presence often indicates soil that does not excessively retain water, allowing good root aeration and preventing saturation.

Tolerance to Nutrient-poor to Moderately Nutrient-rich Soils:
Bellis perennis can develop in soils poor to moderately rich in nutrients. Their presence can indicate soil that tolerates a certain variability in terms of fertility, often found in lawns, meadows, and fallow lands. They also indicate decalcification, and iron loss because of soil erosion.

Preference for Neutral to Slightly Alkaline Soils:
Daisies prefer soils with a pH ranging from slightly alkaline to neutral, typically between 6.0 and 7.5. Their presence may indicate soil with a stable pH, suitable for many meadow and lawn plants.

Preference for Sunny to Partially Shaded Environments:
Bellis perennis grows in environments well-exposed to sunlight or occasionally in partial shade. Their presence can indicate open or semi-open areas.

Tolerance to Frequent Mowing and Trampling:
Daisies resist grazing, frequent mowing, and trampling, making them good indicators of lawns and meadows maintained by humans, overgrazed areas, or regularly trampled zones.

BERBERIS VULGARIS (BARBERRY)

WELL-DRAINED SOILS
THIN SOILS
BASE-RICH SOILS
POOR SOILS
PH OF 6.5 TO 8.0
SUNNY ENVIRONMENTS
DISTURBED HABITATS

Preference for Well-drained Soils:
Barberry thrives in well-drained soils. Its presence often indicates rocky soil that allows good root aeration, or even areas with minimal soil.

Tolerance to Nutrient-poor Soils:
This plant can develop in base-rich but nutrients-poor soils. Its presence indicates soil that is not fertile, often found on slopes, wastelands, and roadsides. It indicates soil rich in bases or alkaline conditions.

Preference for Neutral to Alkaline Soils:
Berberis vulgaris prefers soils with a pH ranging from neutral to slightly alkaline, typically between 6.5 and 8.0. Its presence may indicate soil with a stable, often slightly basic pH.

Preference for Sunny Environments:
Barberry grows well in sunny to partially shaded environments. Its presence can indicate open or semi-open areas with good light exposure.

Tolerance to Disturbance:
Berberis vulgaris is relatively tolerant of disturbance and can colonize disturbed areas. Its presence may indicate habitats that have undergone natural or human disturbances, such as slopes and roadsides.

BETULA PENDULA (SILVER BIRCH)

WIDE VARIETY OF SOIL TYPES
POOR SOILS DEFICIENT IN CALCIUM AND BASES
ACIDIC SOIL PH
SUNNY ENVIRONMENTS
LITTLE COMPETITION
MODERATE DROUGHT TOLERANCE.
DISTURBED HABITATS UNDERGOING NATURAL REGENERATION.

Preference for Dry to Peaty Soils:

Silver birch thrives in dry to peaty soils. Its presence often indicates poor soil but it tolerates a wide variety of moisture quantity and soil types.

Tolerance to Nutrient-poor Soils:

This pioneer species can grow in soils that are relatively poor in organic matter and nutrients. Its presence indicates a base and calcium deficiency.

Preference for Acidic to Neutral Soils:

Betula pendula prefers soils with a pH ranging from acidic to neutral, typically between 4.0 and 7.0. Its presence may indicate soil with a stable, often acidic pH.

Preference for Sunny Environments:

Silver birch grows in environments well-exposed to sunlight. Its presence can indicate open areas with no competition and full light exposure.

Tolerance to Moderate Drought:

Betula pendula is relatively drought-resistant, although it prefers moist conditions. Its presence can indicate occasional periods of drought.

Indication of Disturbance and Regeneration:

Silver birch is often found in disturbed habitats, such as clearings, wastelands, and reforestation areas. It is a pioneer species that can indicate natural regeneration after a disturbance.

BETULA PUBESCENS (DOWNY BIRCH)

WATER-SATURATED SOILS
POOR SOILS
BASE-POOR SOILS
PH OF 4.5 TO 6.5
SUNNY ENVIRONMENTS
DISTURBED HABITATS

Preference for Moist to Waterlogged Soils:

Downy birch thrives in moist to waterlogged soils, often found in marshes, bogs, and wet meadows. Its presence generally indicates water-saturated, poorly drained soil. It can also be an indicator of a spring resurgence. In some cases, it adapts to dry soils.

Tolerance to Poor to Moderately Nutrient-rich Soils:

This species can develop in soils that are poor in bases and poor to moderately rich in nutrients, often on acidic or peaty soils. Its presence may indicate soil that is not particularly fertile but is well-suited to the typical vegetation of wet environments.

Preference for Acidic to Neutral Soils:

Betula pubescens prefers soils with a pH ranging from slightly acidic to neutral, i.e., a pH range of 4.5 to 6.5.

Preference for Partially Shaded to Sunny Environments:

Downy birch thrives in partially shaded to sunny environments. It is often found in clearings, pond edges, forest borders, and open areas in wet environments.

Tolerance to Disturbances:

Betula pubescens is relatively tolerant to disturbances and is a pioneer tree that quickly colonizes disturbed areas, particularly in wet zones. Its presence may indicate a habitat that has undergone natural or human disturbances, such as logging or changes in local hydrology.

BROMUS INERMIS (SMOOTH BROME)

WELL-DRAINED SOILS
MODERATELY RICH SOILS
PH ABOVE 7.5.
SUNNY ENVIRONMENTS
PERIODS OF LOW WATER AVAILABILITY
DISTURBED HABITAT.

Preference for Well-drained to Moderately Moist Soils:

Smooth brome thrives in well-drained to moderately moist soils. Its presence often indicates soil that retains enough moisture without being waterlogged.

Tolerance to Moderately Nutrient-rich Soils:

This plant prefers soils with low organic matter but a sufficient amount of nutrients. Its presence can indicate soil of medium fertility, often found in meadows, roadsides, and cultivated lands.

Preference for Alkaline Soils:

Bromus inermis prefers alkaline soils, with a stable pH above 7.5.

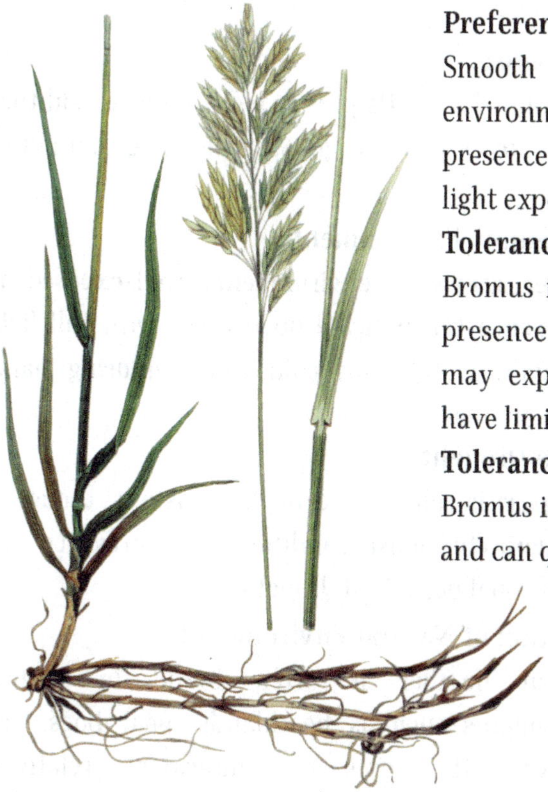

Preference for Sunny Environments:

Smooth brome grows well in environments with full sun exposure. Its presence indicates open areas with full light exposure.

Tolerance to Drought:

Bromus inermis is drought-resistant. Its presence can indicate environments that may experience periods of drought or have limited water availability.

Tolerance to Disturbance:

Bromus inermis is tolerant of disturbance and can quickly colonize disturbed areas. Its presence can indicate habitats that have undergone natural or human disturbances, such as agricultural lands and roadsides.

CALLUNA VULGARIS (COMMON HEATHER)

VERY ACIDIC SOILS
PH OF 3.5 TO 5.5
POOR SOILS
WELL-DRAINED SOILS, OFTEN SANDY OR PEATY
SUNNY ENVIRONMENTS
COLD CLIMATES
TOLERANCE TO DROUGHT
UNDISTURBED NATURAL HABITATS

Preference for Acidic Soils:

Common heather thrives in acidic soils and in conditions of total decalcification. Its presence often indicates soil with a low pH, generally between 3.5 and 5.5. Acidic soils are typical of heathlands and peat bogs.

Tolerance to Nutrient-poor Soils:

This plant can develop in soils poor in nutrients and clay, and sometimes with a complete absence of bases. It may indicate the presence of peat (fossilized plant organic matter) and plant organic matter in a state of fossilization and thus not available.

Preference for Well-drained Soils:

Calluna vulgaris prefers well-drained soils. Its presence can indicate soil that avoids water saturation, often sandy or peaty. It grows in peat bogs but not in waterlogged areas.

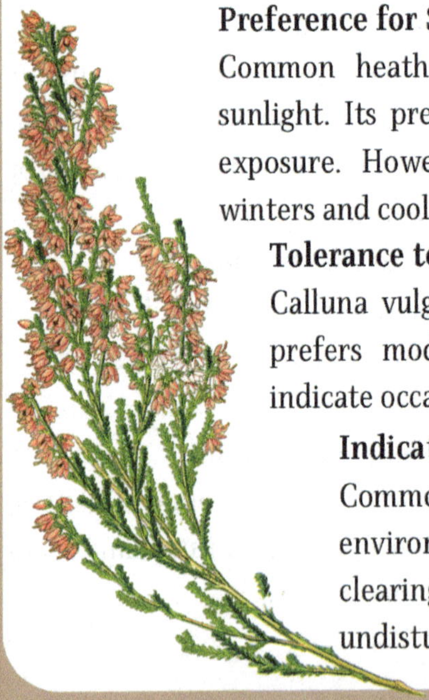

Preference for Sunny Environments:

Common heather grows in environments well-exposed to sunlight. Its presence can indicate open areas with full light exposure. However, it tolerates cold well, enduring harsh winters and cool summers.

Tolerance to Drought:

Calluna vulgaris is relatively drought-resistant, though it prefers moderately moist conditions. Its presence can indicate occasional periods of drought

Indication of Natural Environments:

Common heather is often found in natural environments such as heathlands, peat bogs, and clearings. Its presence indicates relatively undisturbed natural habitats.

CALTHA PALUSTRIS (MARSH MARIGOLD)

WET TO WATERLOGGED SOILS
SOILS SATURATED WITH ORGANIC MATTER
PH OF 5.5 TO 7.0
COOL AND SHADED ENVIRONMENTS
FREQUENTLY FLOODED AREAS
SPRING RESURGENCE

Preference for Wet to Waterlogged Soils:

Marsh marigold thrives in wet to waterlogged soils. Its presence often indicates soil that retains significant moisture, commonly found in floodplains, marshes, and riverbanks.

Tolerance to Soils Saturated With Organic Matter:

This plant prefers soils that are very rich in organic matter and nutrients. Its presence indicates an accumulation of organic matter and a saturated soil as well as the creation of gley (reduction of metal oxides)

Preference for Neutral to Slightly Acidic Soils:

Caltha palustris prefers soils with a pH ranging from slightly acidic to neutral, typically between 5.5 and 7.0.

Preference for Partially Shaded Environments:

Marsh marigold grows well in partially shaded environments. Its presence indicates semi-open areas with a cool climate.

Tolerance to Periodic Flooding:

Caltha palustris is tolerant of periodic flooding, making it an indicator species for frequently flooded areas, near acidic streams, or with a high water table. It is a good indicator for finding a spring resurgence.

CAMPANULA ROTUNDIFOLIA (HAREBELL)

WELL-DRAINED SOILS
POOR SOILS
THIN AND ROCKY SOILS
PH OF 5.0 TO 7.0
DISTURBED ENVIRONMENTS

Preference for Well-drained to Moderately Moist Soils:

Harebell thrives in well-drained soils that are not completely dry.

Tolerance to Nutrient-poor Soils:

This plant indicates soil poor in both animal and plant organic matter and nutrients. The soil lacks structure and clay, often being very thin or rocky. Its presence can indicate soil that is not particularly fertile, commonly found in dry meadows, heaths, slopes, and rocky areas. This plant can adapt well to a wide variety of soils.

Preference for Slightly Acidic to Neutral Soils:

Campanula rotundifolia prefers soils with a pH ranging from acidic to neutral, typically between 5.0 and 7.0. Its presence may indicate soil with a stable pH, suitable for many plants in open habitats.

Preference for Sunny to Partially Shaded Environments:

Harebell grows well in sunny to partially shaded environments, often in a cooler climate.

Tolerance to Disturbed Environments:

Campanula rotundifolia is tolerant of disturbances and can quickly colonize disturbed areas. Its presence may indicate habitats that have undergone natural or human disturbances, such as slopes and rocky areas.

CAPSELLA BURSA-PASTORIS (SHEPHERD'S PURSE)

SOILS WITH SIGNIFICANT WATER VARIATION
NUTRIENT-POOR TO NUTRIENT-RICH SOILS
PH OF 6.0 TO 8.0
SUNNY ENVIRONMENTS
DISTURBED HABITATS

Indicator of Significant Water Variations:

Shepherd's purse thrives in soils where moisture varies greatly.

Tolerance to Nutrient-poor to Nutrient-rich Soils:

This plant develops in loamy or sandy soils, from poor to nutrient-rich. Its presence can indicate soil whose fertility, like moisture, varies. The soil is rich in bases, and the presence of this plant may indicate a blockage in the availability of potassium and phosphorus, often due to compaction.

Preference for Neutral to Slightly Alkaline Soils:

Capsella bursa-pastoris prefers soils with a pH ranging from neutral to slightly alkaline, typically between 6.0 and 8.0. Its presence may indicate soil with a stable and often slightly basic pH.

Preference for Sunny to Partially Shaded Environments:

Shepherd's purse grows well in sunny environments. It can be found in open areas.

Tolerance to Disturbance:

Capsella bursa-pastoris is very tolerant of disturbance and various variations, and can quickly colonize disturbed areas. Its presence can indicate habitats that have undergone natural or human disturbances.

CARDAMINE PRATENSIS (CUCKOOFLOWER)

WET TO WATERLOGGED SOILS
SOIL SATURATED WITH ORGANIC MATTER
PH OF 5.5 TO 7.5
SUNNY TO PARTIALLY SHADED ENVIRONMENTS
PERIODIC FLOODING, SPRING RESURGENCE

Preference for Wet to Waterlogged Soils:
Cuckooflower thrives in wet to waterlogged soils. Its presence often indicates soil that retains moisture a lot, commonly found in wet meadows, marshes, and riverbanks.

Tolerance to Soils Rich in Organic Matter:
This plant prefers soils rich in organic matter and nutrients. Its presence indicates a soil saturated with organic matter.

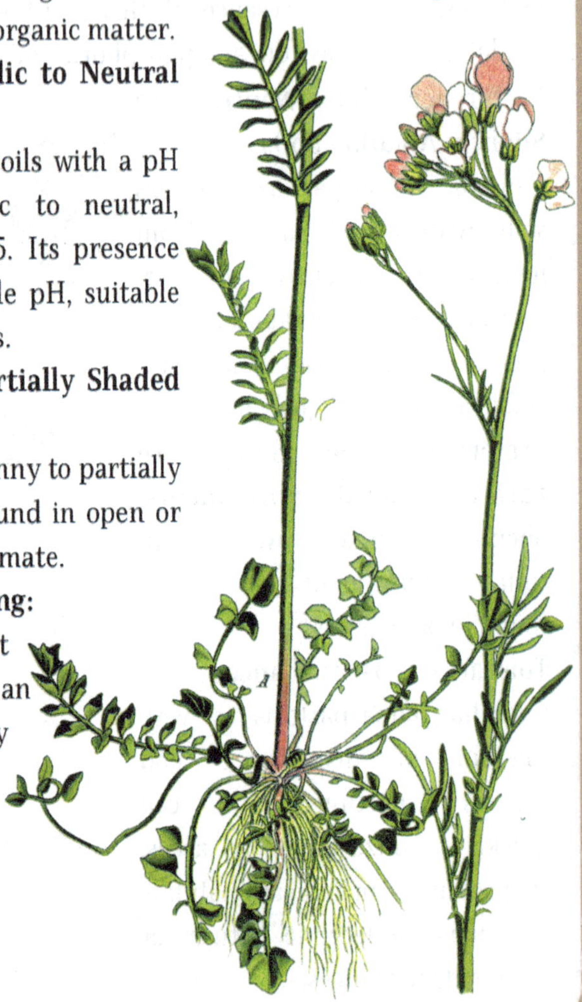

Preference for Slightly Acidic to Neutral Soils:
Cardamine pratensis prefers soils with a pH ranging from slightly acidic to neutral, typically between 5.5 and 7.5. Its presence may indicate soil with a stable pH, suitable for many plants in wet habitats.

Preference for Sunny to Partially Shaded Environments:
Cuckooflower grows well in sunny to partially shaded environments. It is found in open or semi-open areas with a cool climate.

Tolerance to Periodic Flooding:
Cardamine pratensis is tolerant of periodic flooding, making it an indicator species for frequently flooded areas or areas with a high water table, potentially indicating a spring resurgence.

CARLINA ACAULIS (DWARF CARLINE)

WELL-DRAINED TO DRY SOILS
POOR SOILS
BASE-RICH SOILS
PH OF 7.0 TO 8.0
SUNNY ENVIRONMENTS
PERIODS OF LOW WATER AVAILABILITY
MOUNTAINOUS CLIMATE

Preference for Well-drained to Dry Soils:

Dwarf carline thistle thrives in well-drained to dry soils. Its presence often indicates soil that does not retain water, allowing good root aeration.

Tolerance to Poor Soils:

This plant can develop in soils poor in organic matter but with sufficient nutrient availability. Its presence can indicate very thin, rocky soil, rich in bases or with a high pH, and phosphorus blockage.

Preference for Neutral to Alkaline Soils:

Carlina acaulis prefers soils with a stable and alkaline pH, typically ranging from 7.0 to 8.0 or even higher.

Preference for Sunny Environments:

Dwarf carline thistle grows well in environments well-exposed to sunlight. Its presence can indicate open areas with full light exposure.

Tolerance to Drought:

Carlina acaulis is drought-resistant. Its presence can indicate environments that may experience periods of drought or have low water availability.

Indication of Temperate to Mountainous Climatic Conditions:

Dwarf carline thistle is common in temperate to mountainous climatic zones, tolerating cold winters and cool to moderately warm summers. Its presence can indicate a temperate to mountainous climate with distinct seasons. In the mountains, it indicates rain a few hours before it appears by closing its flower, making it a valuable natural barometer. Interestingly, this phenomenon sometimes continues when the plant is dried.

CARPINUS BETULUS (COMMON HORNBEAM)

WELL-DRAINED TO MODERATELY MOIST SOILS
NUTRIENT-RICH SOILS
PH OF 6.0 TO 7.5
PARTIALLY SHADED ENVIRONMENTS
STABLE ECOSYSTEM

Preference for Well-drained to Moderately Moist Soils:
The presence of common hornbeam indicates soil that avoids water saturation while retaining sufficient moisture. Generally, this tree does not like any extremes.

Tolerance to Nutrient-rich Soils:
This species prefers soils rich in available nutrients. Its presence can indicate fertile soil with adequate organic matter content.

Preference for Neutral Soils:
Carpinus betulus prefers soils with a neutral pH, typically between 6 and 7.5. Its presence may indicate soil with a stable pH, suitable for many forest plants. It tolerates limestone relatively well.

Preference for Partially Shaded Environments:
Common hornbeam grows well in partially shaded environments, often found in undergrowth or forest edges. Its presence can indicate moderate light exposure and a cool, moderate climate.

Tolerance to Moderate Drought:
It can tolerate periods of moderate drought.

Indication of Mature Forests:
Common hornbeam is often found in mature, well-established forests where it forms an understory, indicating a relatively stable and undisturbed ecosystem. It thrives in symbiosis with oak and, as a relatively short tree, tolerates being grazed by animals or pruned.

CENTAUREA CYANUS (CORNFLOWER)

WELL-DRAINED TO MODERATELY MOIST SOILS
HYDRIC CONTRASTS
MODERATELY RICH SOILS
PH OF 5.0 TO 7.5
RECENT INCREASE IN PH
SUNNY ENVIRONMENTS
DISTURBED HABITATS

Preference for Well-drained to Moderately Moist Soils:
Cornflower thrives in well-drained to moderately moist soils. Its presence often indicates soil that retains sufficient moisture without being constantly waterlogged. It is very tolerant of moisture variations and can indicate significant hydric contrasts.

Tolerance to Poor to Moderately Nutrient-rich Soils:
This plant can develop in soils that are poor to moderately rich in nutrients. Its presence may indicate soil that is not particularly fertile, often found in cultivated fields, fallows, and roadsides.

Preference for Neutral to Slightly Acidic Soils:
Centaurea cyanus prefers soils with a pH ranging from slightly acidic to neutral, i.e., a pH range of 5.0 to 7.5. It often indicates a recent increase in soil pH, especially if it grows alongside poppies. The pH increase can be due to liming, sheep grazing, an increase in temperature, or other reasons.

Preference for Sunny Environments:
Cornflower thrives in well-exposed sunny environments. Its presence can indicate open areas with full light exposure.

Tolerance to Disturbances:
Centaurea cyanus is relatively tolerant of disturbance and can quickly colonize disturbed areas. Its presence indicates a habitat that has undergone natural or human disturbances, such as agricultural lands, fallows, and roadsides.

CENTAUREA JACEA (BROWN KNAPWEED)

STRONG WATER CONTRASTS
ORGANIC MATTER IMBALANCE
PH OF 5.5 TO 7.0
SUNNY ENVIRONMENTS
DISTURBED HABITAT

A Sign of Water Contrasts:

Centaurea jacea often indicates soil that undergoes strong water contrasts, with periods of drought in summer and very wet periods in winter.

Tolerance to Imbalance:

This plant prefers soils rich in carbon-based plant organic matter and bases, and poor in animal organic matter. It indicates a lack of nitrogen and potassium.

Preference for Neutral Soils:

Centaurea jacea prefers soils with a pH ranging from slightly acidic to neutral, with a pH range of 5.5 to 7.0.

Preference for Sunny Environments:

Centaurea jacea thrives in sunny environments. Its presence can indicate open or semi-open areas with good light exposure and a cool climate.

Tolerance to Disturbance:

Centaurea jacea is relatively tolerant to disturbance and can quickly colonize disturbed areas. Its presence may indicate a habitat that has undergone natural or human disturbances.

CENTAURIUM ERYTHRAEA (COMMON CENTAURY)

MODERATELY MOIST SOILS
BASE-RICH SOILS
MEDIUM FERTILITY SOILS
PH OF 6 TO 7.5
SUNNY ENVIRONMENTS
DISTURBED HABITATS

Preference for Well-drained to Moderately Moist Soils:
Common centaury thrives in soils with strong hydric contrasts. The soil may be waterlogged in winter and prone to hydromorphisms, and conversely dry in summer.

Tolerance to Moderately Nutrient-rich Soils:
This plant can develop in base-rich soils and moderately rich in available nutrients. Its presence can indicate soil with moderate fertility.

Preference for Neutral to Basic Soils:
Centaurium erythraea prefers soils with a stable pH ranging from neutral to basic, typically between 6 and 7.5.

Preference for Sunny to Partially Shaded Environments:
Common centaury grows well in sunny environments with a cool climate.

Tolerance to Disturbance:
Centaurium erythraea is relatively tolerant of disturbance and can quickly colonize disturbed areas. Its presence can indicate habitats that have undergone natural or human disturbances, such as fallow lands and roadsides.

CHELIDONIUM MAJUS (GREATER CELANDINE)

VERY NUTRIENT-RICH SOILS
WELL-DRAINED TO MODERATELY MOIST SOILS
PH FROM 6.5 TO 8.0
OPEN OR SEMI-OPEN AREAS
DISTURBED HABITATS

Preference for Nutrient-Rich Soils:

Chelidonium majus thrives in soils very rich in available nutrients, particularly bases and nitrogen. It seeks nitrates and often grows in soils rich in organic matter of animal origin. It can also grow in very thin soils as long as nutrients are available.

Tolerance to Well-Drained to Moderately Moist Soils:

Chelidonium majus prefers well-drained to moderately moist soils.

Preference for Neutral to Slightly Alkaline Soils:

Chelidonium majus prefers soils with a pH ranging from neutral to slightly alkaline, with a pH range of 6.5 to 8.0.

Preference for Partially Shaded to Sunny Environments:

Chelidonium majus thrives in partially shaded to sunny environments. Its presence can indicate open or semi-open areas with good light exposure, in a cool climate.

Indication of Disturbance and Ruderalization:

This plant is often found in disturbed habitats and can indicate recent or historical habitat disturbance.

CHENOPODIUM ALBUM (GOOSEFOOT)

WATER CONTRASTS
NITROGEN-RICH SOILS
DISTURBED AREAS
FULL SUN EXPOSURE
SOIL PH FROM 5.5 TO 8.0
POSSIBLE HEAVY METAL CONTAMINATION
MODERATELY HIGH SOIL SALINITY

Tolerance to a Wide Range of Soils:

Chenopodium album indicates an excess of non-composted animal organic matter (nitrogen) in soil with strong water contrasts. It withstands periods of drought.

Indication of Nutrient-Rich Soils:

This plant thrives in soils rich in available nutrients, particularly nitrogen. Its presence can show areas with fertile and well-fertilized soils, often due to agricultural runoff, manure application, or environments favored by animals, particularly chickens or birds whose droppings are very rich in nitrogen.

Resilience to Disturbances:

Chenopodium album is commonly found in disturbed areas, such as agricultural fields, gardens, and wastelands. Its presence suggests recent soil disturbance or human activity.

Sensitivity to Shading:

Preferring sunny environments, Chenopodium album is typically found in open habitats such as fields, roadsides, and wastelands. Its presence indicates full sun exposure or conditions of low shading.

Slightly acidic to slightly alkaline pH:

Chenopodium album can grow in a range of soil pH, from slightly acidic to slightly alkaline (pH 5.5 to 8.0). Its presence suggests a soil environment that is neither extremely acidic nor extremely alkaline.

Tolerance to Heavy Metals:

This plant has shown some tolerance to heavy metals, making it an indicator of soils potentially contaminated by these substances, particularly in urban or industrial areas.

Indication of High Soil Salinity:

Chenopodium album can tolerate moderately high levels of soil salinity. Its presence can indicate saline soils, often due to the natural composition of the soil or irrigation in hot weather.

CHRYSANTHEMUM LEUCANTHEMUM (COMMON DAISY)

WELL-DRAINED SOILS
MODERATELY NUTRIENT-RICH SOILS
PH OF 5.5 TO 7.0
SUNNY TO PARTIALLY SHADED ENVIRONMENTS
TOLERANCE TO DISTURBANCE

Preference for Well-Drained to Moderately Moist Soils:

The common daisy thrives in well-drained to moderately moist soils. Its presence often indicates soil that retains enough moisture without being constantly waterlogged.

Tolerance to Poor to Moderately Rich Nutrient Soils:

This plant can grow in soils that are poor to moderately rich in nutrients and organic matter. The soil is often rich in bases.

Preference for Neutral to Slightly Basic Soils:

Chrysanthemum leucanthemum prefers soils with a pH ranging from neutral to slightly basic, with a pH range of 5.5 to 7.5. Its presence can indicate soil with a stable pH, suitable for many plants in open and semi-open environments.

Preference for Sunny Environments:

The common daisy thrives in sunny environments. Its presence can indicate open or semi-open areas with good light exposure in a cool climate.

Tolerance to Disturbance:

Chrysanthemum leucanthemum is relatively tolerant to disturbance and can quickly colonize disturbed areas.

Its presence may indicate a habitat that has undergone natural or human disturbances, such as meadows and roadsides.

CICHORIUM INTYBUS (WILD CHICORY)

WELL-DRAINED SOILS
SOILS POOR IN ORGANIC MATTER
SOILS RICH IN BASES
PH OF 6.5 TO 8.0
SUNNY ENVIRONMENTS
TOLERANCE TO DROUGHT
DISTURBED HABITATS

Preference for Well-Drained to Moderately Moist Soils:
Wild chicory thrives in well-drained to moderately moist soils. Its presence often indicates soil that avoids water saturation while retaining sufficient moisture.

Tolerance to Poor to Moderately Nutrient-Rich Soils:
This plant can develop in soils poor in organic matter but rich in nutrients, especially bases. It indicates soil compaction and settling.

Preference for Neutral to Slightly Alkaline Soils:
Cichorium intybus prefers soils with an alkaline pH, ranging from 6.5 to 8.0. It grows poorly in acidic soils.

Preference for Sunny Environments:
Wild chicory grows in well-exposed sunny environments. Its presence can indicate open areas with constant light exposure in a warm temperate climate.

Tolerance to Drought:
Cichorium intybus is relatively drought-resistant, making it a good indicator of environments that may undergo periods of drought or have low water availability.

Indication of Disturbance:
This plant is often found in disturbed habitats, such as roadsides, wastelands, and vacant lots. Its presence may indicate habitat disturbance, often related to human activity.

CICUTA VIROSA (WATER HEMLOCK)

VERY MOIST TO WATERLOGGED SOILS
CLAY SOIL
SOILS RICH IN ORGANIC MATTER
PH OF 6.5 TO 8.0
SUNNY TO PARTIALLY SHADED ENVIRONMENTS

Preference for Very Moist to Waterlogged Soils:

Water hemlock thrives in very moist to waterlogged soils. Its presence often indicates soil with high or periodic water saturation, commonly found in marshes, ditches, and along water bodies.

Tolerance to Soils Rich in Organic Matter:

This plant prefers clay soils that are extremely rich in organic matter and nutrients. Its presence can indicate fertile soil, often with an accumulation of organic matter due to the decomposition of aquatic vegetation.

Preference for Neutral to Slightly Alkaline Soils:

Cicuta virosa prefers soils with an alkaline pH, ranging from 6.5 to 8.

Its presence can indicate soil with a stable pH, often found in aquatic or semi-aquatic habitats.

Preference for Sunny to Partially Shaded Environments:

Water hemlock grows in well-exposed sunny or partially shaded environments in cool climates. Its presence can indicate open or semi-open areas, often near water bodies.

CIRSIUM ARVENSE (FIELD THISTLE)

WELL-DRAINED TO MODERATELY MOIST SOILS
EXCESS UNCOMPOSED MANURE
OVERGRAZING
PH OF 6.0 TO 7.5
SUNNY ENVIRONMENTS
SEVERELY DISTURBED HABITAT

Preference for Well-Drained to Moderately Moist Soils:
Field thistle thrives in well-drained to moderately moist soils. Its presence often indicates soil that retains sufficient moisture without being constantly waterlogged.

Tolerance to Nutrient-Rich Soils:
This plant prefers nutrient-rich soils. Its presence indicates a saturation of organic matter, often due to excessive application of uncomposed manure or overgrazing. The soil phosphorus may become locked.

Preference for Neutral to Slightly Alkaline Soils:
Cirsium arvense prefers soils with a pH ranging from neutral to slightly alkaline, within a range of 6.0 to 7.5. Its presence can indicate soil with a stable and often slightly basic pH.

Preference for Sunny Environments:
Field thistle grows well in sunny environments in a cool climate.

Tolerance to Disturbance:
Cirsium arvense is highly tolerant to disturbance and can quickly colonize disturbed areas. Its presence may indicate a habitat that has undergone significant natural or human disturbances.

CLEMATIS VITALBA (HEDGE CLEMATIS)

WELL-DRAINED SOILS
SOILS RICH IN ORGANIC MATTER AND NUTRIENTS
PH OF 6.5 TO 8.0
OPEN OR SEMI-OPEN AREAS
DISTURBED HABITATS

Preference for Well-Drained to Moderately Moist Soils:

Hedge clematis thrives in well-drained to moderately moist soils. Its presence often indicates soil that does not retain excessive water, allowing good aeration of the roots.

Tolerance to Nutrient-Rich Soils:

This plant prefers soils rich in organic matter and nutrients, particularly bases and nitrogen. It thrives at forest edges, cultivated areas, and even nitrogen-polluted urban environments.

Preference for Neutral to Alkaline Soils:

Clematis vitalba prefers soils with a pH ranging from neutral to slightly alkaline, within a range of 6.5 to 8.0. Its presence can indicate soil with a stable, often slightly basic, pH.

Preference for Sunny to Partially Shaded Environments:

Hedge clematis grows in well-exposed sunny or partially shaded environments. Its presence can indicate open or semi-open areas with good light exposure.

Indication of Disturbance:

This plant is often found in disturbed habitats, such as roadsides, hedges, and abandoned gardens. Its presence can indicate habitat disturbance, often related to human activity.

CLINOPODIUM VULGARE (COMMON SAVORY)

WELL-DRAINED SOILS
POOR SOILS
PH OF 6.5 TO 8.0
SUNNY ENVIRONMENTS
DROUGHT TOLERANCE

Preference for Well-drained Soils:
Common savory thrives in well-drained soils. Its presence often indicates a soil that does not retain excessive water, allowing good aeration for the roots.
Tolerance for Poor to Moderately Nutrient-rich soils:
This plant can grow in soils poor in organic matter and moderately rich in available nutrients. Its presence indicates soil that is not particularly fertile, often found in dry meadows, roadside edges, and calcareous lawns.

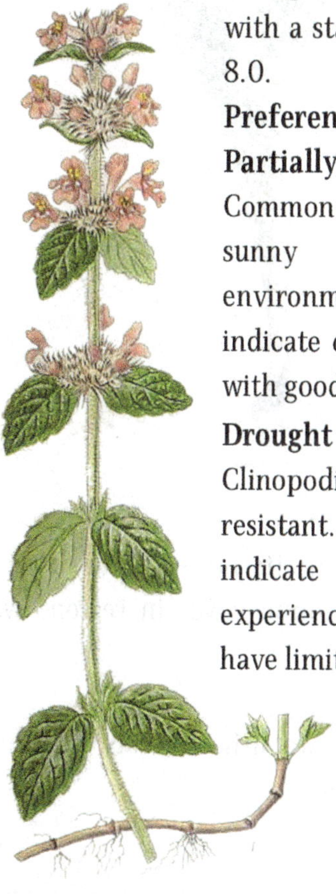

Preference for Alkaline Soils:
Clinopodium vulgare prefers soils with a stable pH between 6.5 and 8.0.
Preference for Sunny to Partially Shaded Environments:
Common savory grows well in sunny to partially shaded environments. Its presence may indicate open or semi-open areas with good light exposure.
Drought Tolerance:
Clinopodium vulgare is drought-resistant. Its presence may indicate environments that can experience periods of drought or have limited water availability.

CONVALLARIA MAJALIS (LILY OF THE VALLEY)

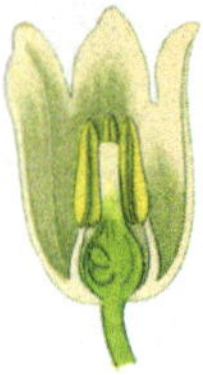

WELL-DRAINED SOILS
SOILS BALANCED IN ORGANIC MATTER
WIDE RANGE OF SOIL PH
SHADED ENVIRONMENTS
STABLE ECOSYSTEMS
MATURE AND ANCIENT FORESTS

Preference for Well-drained to Moderately Moist Soils:
Lily of the valley thrives in well-drained to moderately moist soils. Its presence often indicates soil that retains enough moisture while allowing good aeration for the roots.

Well Balanced Soils:
This plant prefers soils moderately rich in organic matter, bases and nutrients. Its presence can indicate a well-balanced, healthy, and very stable forest floor soil type.

Preference for Slightly Acidic to Neutral Soils:
Convallaria majalis prefers soils with a pH range from slightly acidic to neutral, specifically a pH range of 5.0 to 7.0. Its presence can indicate a stable pH suitable for many forest plants but it is tolerant to a wide range of soil pH.

Preference for Partially Shaded Environments:
Lily of the valley grows in shaded environments. Its presence can indicate undergrowth or forest edges with dense vegetation cover in regions with relatively warm summer-weather..

Indication of Mature Forests:
Lily of the valley is often found in mature, well-established forests, indicating an old, stable, and minimally disturbed ecosystem.

CORNUS SANGUINEA (COMMON DOGWOOD)

WELL-DRAINED SOILS
MODERATELY NUTRIENT-RICH SOILS
PF OF 7.0 TO 8.0
PARTIALLY SHADED ENVIRONMENTS
DISTURBED HABITATS

Preference for Well-drained to Moderately Moist Soils:

The common dogwood thrives in well-drained to moderately moist soils. Its presence often indicates soil that retains sufficient moisture without being constantly waterlogged.

Soils Rich in Organic Matter:

This plant prefers base-rich soils moderately rich in organic matter and nutrients. Its presence can indicate relatively fertile soil, often found in forests, forest edges, and hedgerows.

Preference for Alkaline Soils:

Cornus sanguinea prefers soils with a pH range from neutral to slightly alkaline, specifically a pH range of 7.0 to 8.0. Its presence can indicate stable, basic soil pH.

Preference for Sunny to Partially Shaded Environments:

The common dogwood grows well in partially shaded environments. Its presence may indicate open or semi-open areas with good light exposure.

Tolerance to Disturbance:

Cornus sanguinea is relatively tolerant of disturbance and can colonize disturbed areas. Its presence may indicate habitats that have undergone natural or human disturbances, such as hedgerows and roadside edges.

CORYLUS AVELLANA (COMMON HAZEL)

SOILS THAT RETAIN LITTLE WATER
SOILS POOR IN ANIMAL ORGANIC MATTER
PH OF 5.0 TO 7.0
PARTIALLY SHADED ENVIRONMENTS
PIONEER AND EDGE SPECIES

Preference for Well-drained to Moderately Moist Soils:

The common hazel thrives in well-drained to moderately moist soils. Its presence often indicates soil that does not retain moisture well and dries out easily.

Soil Evolving Towards Forest:

This tree grows in base-rich soils that are evolving towards a forest state. Such soils are rich in carbon-based organic matter and poor in animal organic matter.

Preference for Neutral to Slightly Acidic Soils:

Corylus avellana prefers soils with a pH range from slightly acidic to neutral, specifically a pH range of 5.0 to 7.0. Its presence can indicate stable soil pH suitable for many forest plants.

Preference for Partially Shaded to Sunny Environments:

The common hazel grows well in partially shaded to sunny environments in cool or even mountainous climates. Its presence may indicate areas with variable vegetation cover, often found at forest edges and in hedgerows.

Recolonization or Forest Edge:

The common hazel is often a pioneer tree after forest clearing but can also be found at the edge of a mature and well-established hillside or mountain forest. It frequently recolonizes abandoned agricultural areas.

CRATAEGUS MONOGYNA (HAWTHORN)

WELL-DRAINED SOILS
TOLERANCE OF DIFFERENT LEVELS OF FERTILITY
PH OF 6.0 TO 8.0
SEMI-OPEN AREAS
STABLE OR DISTURBED HABITATS

Preference for Well-drained to Moderately Moist Soils:

The common hawthorn thrives in well-drained to moderately moist soils. Its presence often indicates soil that retains sufficient moisture without becoming waterlogged.

Tolerance for Poor to Moderately Nutrient-rich Soils:

This plant can develop in base-rich soils that range from poor to moderately nutrient-rich. Its presence generally indicates an excess of carbon-based plant organic matter and a deficiency in animal organic matter. It is, however, tolerant of some variability in fertility, often found in hedgerows, meadows, and forest edges.

Preference for Neutral to Slightly Alkaline Soils:

Crataegus monogyna prefers soils with a pH range from neutral to slightly alkaline, specifically a pH range of 6.0 to 8.0. Its presence can indicate stable soil pH, often slightly basic.

Preference for Sunny to Partially Shaded Environments:

The common hawthorn grows well in sunny to partially shaded environments. Its presence may indicate open or semi-open areas with good light exposure.

Indication of Moderate to Low Disturbance:

This plant is often found in disturbed habitats such as fallows, vineyards, hedgerows, roadsides, and meadows. Its presence can indicate moderate habitat disturbance, often related to human activity.

DACTYLIS GLOMERATA (COCK'S-FOOT GRASS)

WELL-DRAINED TO MODERATELY MOIST SOILS
SOILS RICH IN ANIMAL ORGANIC MATTER
PH OF 5.5 TO 7.5
OPEN OR SEMI-OPEN AREAS
DISTURBED HABITATS

Preference for Well-drained to Moderately Moist Soils:

Cock's-foot grass thrives in well-drained to moderately moist soils. Its presence often indicates soil that retains sufficient moisture without being constantly waterlogged.

Tolerance for Moderately Nutrient-rich to Rich Soils:

This plant prefers soils rich in animal organic matter and bases. Its presence can indicate an excess in animal organic matter (nitrogen) and a soil where available nutrients are starting to be blocked, most often because of soil compaction.

Preference for Slightly Acidic to Slightly Basic Soils:

Dactylis glomerata prefers soils with a pH range from slightly acidic to slightly basic, specifically a pH range of 5.5 to 7.5. Its presence can indicate stable soil pH suitable for many plants in open and semi-open environments.

Preference for Sunny to Partially Shaded Environments:

Cock's-foot grass grows well in sunny to partially shaded environments. Its presence may indicate open or semi-open areas with good light exposure.

Tolerance to Disturbance:

Dactylis glomerata is relatively tolerant of disturbance and can quickly colonize disturbed areas. Its presence can indicate habitats that have undergone natural or human disturbances, such as compaction on pastures or roadsides.

DAUCUS CAROTA (WILD CARROT)

SOILS SUBJECT TO WATER CONTRASTS
SOILS POOR IN ORGANIC MATTER
SOILS RICH IN AVAILABLE NUTRIENTS
PH OF 6.0 TO 8.0
SUNNY ENVIRONMENTS
DISTURBED HABITATS

Preference for Well-drained Soils:

Wild carrot thrives in well-drained soils. Its presence often indicates soil rich in bases and subject to water contrasts, such as along roadsides.

Tolerance for Poor to Moderately Nutrient-rich Soils:

This plant can develop in soils poor in fresh organic matter but rich in available nutrients.

Preference for Neutral to Slightly Alkaline Soils:

Daucus carota prefers soils with a pH range from neutral to slightly alkaline, specifically a pH range of 6.0 to 8.0.

Preference for Sunny Environments:

Wild carrot grows in environments well-exposed to sunlight.

Indication of Disturbance:

This plant is often found in disturbed habitats, such as roadsides, abandoned fields, and meadows. Its presence can indicate habitat disturbance, often related to human activity.

DIGITALIS PURPUREA (FOXGLOVE)

WELL-DRAINED SOILS
SOILS WITH AN EXCESS OF PLANT ORGANIC MATTER
PH OF 4.5 TO 7.0
PARTIALLY SHADED TO SUNNY ENVIRONMENTS
ABRUPT TRANSITION FROM SHADOW TO LIGHT
DISTURBED HABITATS
SUDDEN SOIL LOSS

Preference for Well-drained to Moderately Moist Soils:

Foxglove thrives in well-drained to moderately moist soils. Its presence often indicates soil that retains sufficient moisture while allowing good root aeration.

Tolerance for Poor Soils:

This plant can develop in relatively poor soils with an excess of plant organic matter. It is often found on embankments, slopes, and recent clear-cut areas, illustrating soil loss.

Preference for Acidic to Neutral Soils:

Digitalis purpurea prefers soils with a pH range from slightly acidic to neutral, specifically a pH range of 4.5 to 7.0.

Preference for Partially Shaded to Sunny Environments:

Foxglove grows well in partially shaded to sunny environments. Its presence may indicate open or semi-open areas with good light exposure. This plant often appears during a sudden transition from shade to light, typically following forest cutting.

Indication of Disturbance:

This plant is found in recently or constantly disturbed habitats, subject to soil loss.

DRYOPTERIS FILIX-MAS (MALE FERN)

MOIST TO WELL-DRAINED SOILS
HIGH ATMOSPHERIC HUMIDITY
SOILS RICH IN PLANT ORGANIC MATTER
LACK OF ANIMAL ORGANIC MATTER
PH OF 5.0 TO 7.0
COOL, SHADED ENVIRONMENTS

Preference for Moisture-retentive Soils:

Male fern thrives in soils that retain sufficient moisture while allowing good drainage to avoid waterlogging. Its presence is an indicator of significant atmospheric humidity.

High Plant Organic Matter Content:

This plant prefers soils rich in plant organic matter and indicates a lack of animal organic matter.

Preference for Acidic to Neutral Soils:

Dryopteris filix-mas prefers soils with a pH range from slightly acidic to neutral, specifically a pH range of 5.0 to 7.0. Its presence can indicate stable soil pH suitable for many forest plants.

Preference for Shaded to Partially Shaded Environments:

Male fern grows well in shaded to partially shaded environments in cool climates. Its presence may indicate undergrowth or forest edges with dense vegetation cover.

Undisturbed Environments:

Dryopteris filix-mas is often found in undisturbed environments, indicative of an ecosystem evolving towards a forested state.

ECHIUM VULGARE (VIPER'S BUGLOSS)

DRY SOILS
POOR SOILS
PH OF 6.0 TO 8.0
SUNNY ENVIRONMENTS
PERIODS OF LOW WATER AVAILABILITY
DISTURBED HABITATS

Preference for Well-drained to Dry Soils:

Viper's bugloss thrives in well-drained to dry soils. Its presence often indicates soil that does not retain water well.

Tolerance for Poor to Moderately Nutrient-rich Soils:

This plant can develop in very poor soils with minimal organic matter and little to no humus. The soil may still contain available nutrients.

Preference for Neutral to Slightly Alkaline Soils:

Echium vulgare prefers soils with a pH range from neutral to slightly alkaline, specifically a pH range of 6.0 to 8.0. Its presence can indicate stable soil pH, often slightly basic.

Preference for Sunny Environments:

Viper's bugloss grows well in environments with full sun exposure.

Drought Tolerance:

Echium vulgare is drought-resistant. Its presence indicates environments that can experience periods of drought or have poor water availability.

Tolerance to Disturbance:

Echium vulgare is tolerant of disturbance and can quickly colonize disturbed areas. Its presence may indicate habitats with no humus that have undergone natural or human disturbances, such as wastelands and roadsides.

EPILOBIUM ANGUSTIFOLIUM (FIREWEED)

WELL-DRAINED SOILS
SOILS SATURATED WITH PLANT ORGANIC MATTER
NITROGEN-RICH SOILS
PH OF 5.0 TO 7.0
OPEN OR SEMI-OPEN AREAS
DISTURBED HABITATS
RECENT DESTRUCTION OF PLANT MATTER

Preference for Well-drained to Moderately Moist Soils:

Fireweed thrives in well-drained to moderately moist soils. Its presence often indicates soil that retains sufficient moisture without being constantly waterlogged.

Tolerance for Poor to Moderately Nutrient-rich Soils:

This plant can develop in soils saturated with plant organic matter and nitrogen.

Preference for Slightly Acidic to Neutral Soils:

Epilobium angustifolium prefers soils with a pH range from slightly acidic to neutral, specifically a pH range of 5.0 to 7.0, but it tolerates a wide range of soil pH.

Preference for Sunny to Partially Shaded Environments:

Fireweed grows well in sunny to partially shaded environments in cool climates.

Tolerance to Disturbance:

Epilobium angustifolium is very tolerant of disturbance and can quickly colonize disturbed areas. Its presence can indicate habitats that have undergone recent natural or human disturbances, particularly destruction of plant matter such as cleared lands, logging areas, post-fire zones, and roadsides.

EQUISETUM ARVENSE (FIELD HORSETAIL)

MOIST SOILS
DEEP SOILS
SPRING RESURGENCE
SOILS POOR IN ORGANIC MATTER
LOSS OF NUTRIENTS
PH OF 4.5 TO 7.5
SUNNY TO PARTIALLY SHADED ENVIRONMENTS
DISTURBED HABITATS

Preference for Moist to Moderately Moist Soils:

Field horsetail thrives in moist to moderately moist and deep soils. Its presence often indicates soil with good water availability and may signal a spring resurgence or groundwater presence.

Tolerance for Nutrient-poor Soils:

This plant can develop in soils poor in organic matter and losing its nutrients.

Preference for Acidic to Neutral Soils:

Equisetum arvense prefers soils with a pH range from slightly acidic to neutral, specifically a pH range of 4.5 to 7.5.

Preference for Sunny to Partially Shaded Environments:

Field horsetail grows well in sunny to partially shaded environments. Its presence may indicate open or semi-open areas with good light exposure in cool climates.

Indication of Disturbance:

This plant is often found in disturbed habitats, such as alluvial valleys, cultivated fields, and fallows. Its presence can indicate habitat disturbance, often related to water movement or human activity.

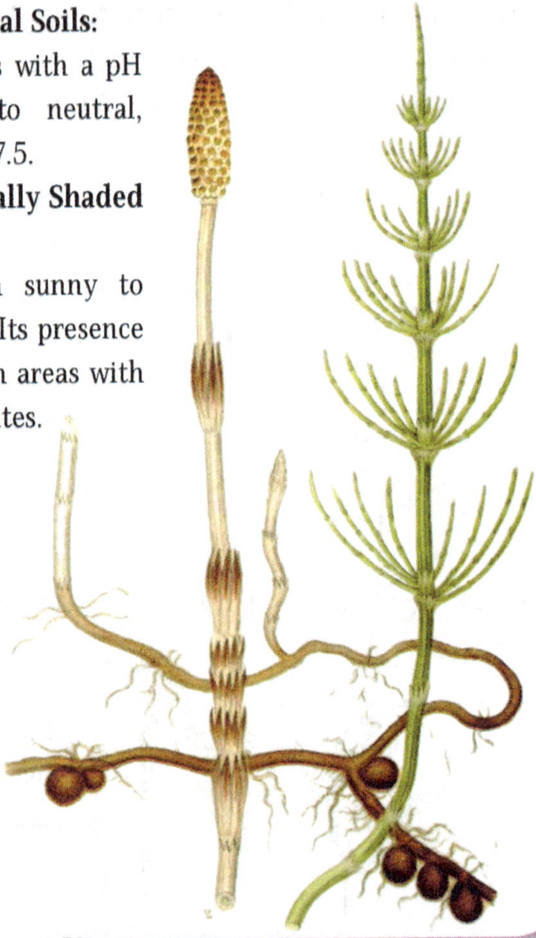

ERYNGIUM MARITIMUM (SEA HOLLY)

SANDY AND UNSTABLE SOILS
POOR SOILS
PH OF 6.5 TO 8.5
PRESENCE OF SALT
SUNNY ENVIRONMENTS

Preference for Sandy Soils:

Sea holly thrives in sandy soils. Its presence often indicates a light-textured, well-drained, and unstable soil typical of dunes and beaches.

Tolerance for Nutrient-poor Soils:

This plant can develop in soils very poor in organic matter and nutrients.

Preference for Neutral to Slightly Alkaline Soils:

Eryngium maritimum prefers soils with a pH range from neutral to slightly alkaline, specifically a pH range of 6.5 to 8.5. Its presence can indicate stable soil pH, often slightly basic.

Tolerance to Salinity:

This plant is tolerant of salinity. Its presence can indicate an environment influenced by sea spray and salt deposits.

Preference for Sunny Environments:

Sea holly grows well in environments with full sun exposure. Its presence indicates open areas with full light exposure.

EUPATORIUM CANNABINUM (HEMP-AGRIMONY)

WATERLOGGED SOILS
BASE-RICH SOILS
SOILS SATURATED WITH PLANT ORGANIC MATTER
PH OF 6.0 TO 8.0
OPEN OR SEMI-OPEN AREAS
DISTURBED HABITATS

Preference for Wet to Waterlogged Soils:

Hemp-agrimony thrives in wet to waterlogged soils. Its presence indicates natural hydromorphism or grazing and/or soil work during wet weather.

Tolerance for Soils Rich in Organic Matter:

This plant indicates base-rich soil that is very rich and saturated with plant organic matter and nutrients. Its presence signifies a wet area with an accumulation of organic matter.

Preference for Neutral to Basic soils:

Eupatorium cannabinum prefers soils with a pH range from neutral to basic, specifically a pH range of 6.0 to 8.0.

Preference for Sunny to Partially Shaded Environments:

Hemp-agrimony grows well in sunny to partially shaded environments. Its presence may indicate open or semi-open areas with good light exposure in cool climates.

Tolerance to Disturbance:

Eupatorium cannabinum is relatively tolerant of disturbance and can quickly colonize disturbed areas. Its presence may indicate habitats that have undergone natural or human disturbances, often involving the destruction of plant matter, such as logging in wet areas.

EUPHORBIA CYPARISSIAS (CYPRESS SPURGE)

DRY SOILS
POOR AND DEGRADED SOILS
EXCESS OF PLANT ORGANIC MATTER
PH OF 6.0 TO 8.0
SUNNY ENVIRONMENTS
DROUGHT TOLERANCE
DISTURBED, TRAMPLED, AND OXYGEN-DEPRIVED HABITATS

Preference for Well-drained Soils:

Cypress spurge thrives in dry soils. Its presence often indicates well-drained conditions.

Tolerance for Eroded Soils:

This plant can develop in base-rich but nutrient-poor soils. Its presence indicates an excess of plant organic matter and degraded, eroded, leached, and sunburned soil that has lost its humus, balance, and structure and is lacking in oxygen.

Preference for Neutral to Slightly Alkaline Soils:

Euphorbia cyparissias prefers soils with a pH range from neutral to slightly alkaline, specifically a pH range of 6.0 to 8.0. Its presence can indicate stable soil pH, often slightly basic.

Preference for Sunny Environments:

Cypress spurge grows well in environments with full sun exposure. Its presence may indicate open areas with full light exposure.

Drought Tolerance:

Euphorbia cyparissias is relatively drought-resistant. Its presence may indicate environments that can experience periods of drought or have low water availability.

Indication of Moderate Disturbance:

This plant is found in disturbed and degraded habitats, often due to human activity. The soil may have been overgrazed, trampled, or worked with heavy machinery.

EUPHRASIA OFFICINALIS (EYEBRIGHT)

DRY SOILS
BASE-RICH SOILS
POOR SOILS
DEFICIENCY IN ANIMAL ORGANIC MATTER
PH OF 5.0 TO 8.0
OPEN OR SEMI-OPEN AREAS
DISTURBED HABITATS

Preference for Well-drained to Moderately Moist Soils:

Eyebright thrives in rather dry soils that retain little moisture. Its presence often indicates well-drained conditions.

Tolerance for Poor to Moderately Nutrient-rich Soils:

This plant can develop in poor to moderately nutrient-rich soils. Its presence may indicate slight saturation with plant organic matter and a deficiency in animal organic matter.

Preference for Neutral to Basic Soils:

Euphrasia officinalis prefers soils with a pH range from slightly acidic to alkaline, specifically a pH range of 5.0 to 8.0.

Preference for Sunny to Partially Shaded Environments:

Eyebright grows well in sunny to partially shaded environments. Its presence may indicate open or semi-open areas with good light exposure.

Indication of Moderate to Low Disturbance:

This plant is often found in disturbed or degraded habitats, such as meadows, pastures, roadsides, and ski slopes in the mountains. Its presence may indicate moderate habitat disturbance, often related to human activity.

FAGUS SYLVATICA (COMMON BEECH)

WELL-DRAINED TO MODERATELY MOIST SOILS
HIGH ATMOSPHERIC HUMIDITY
SOILS RICH IN BASES, ORGANIC MATTER, AND NUTRIENTS
PH OF 4.5 TO 7.5
SHADED TO PARTIALLY SHADED ENVIRONMENTS
STABLE ECOSYSTEM

Preference for Well-drained to Moderately Moist Soils:

The common beech thrives in well-drained to moderately moist soils. Its presence often indicates soil that retains sufficient moisture while allowing good root aeration. This tree signals an environment with high atmospheric humidity.

Preference for Nutrient-rich Soils:

This species prefers soils rich in bases, organic matter, and nutrients.

Preference for Neutral to Slightly Acidic Soils:

Fagus sylvatica prefers soils with a pH range from slightly acidic to neutral, specifically a pH range of 4.5 to 7.5. Its presence can indicate stable soil pH suitable for many forest plants.

Preference for Shaded to Partially Shaded Environments:

The common beech grows well in shaded to partially shaded environments. Its presence may indicate undergrowth or forest edges in cool climates.

Undisturbed Ecosystem:

Fagus sylvatica is often found in mature, well-established forests, indicating a relatively stable and undisturbed ecosystem.

FILIPENDULA ULMARIA (MEADOWSWEET)

WATERLOGGED SOILS
SOILS RICH IN PLANT ORGANIC MATTER
PH OF 5.5 TO 7.0
OPEN OR SEMI-OPEN AREAS
HIGH WATER TABLE

Preference for Wet to Waterlogged Soils:

Meadowsweet thrives in waterlogged soils prone to hydromorphism. The soil is oxygen-deprived, leading to iron reduction and gley formation. Organic matter ceases to decompose and transforms into peat. It can also develop on soils that are not constantly waterlogged but experience periodic flooding, It can indicate a potential spring resurgence.

Tolerance for Soils Rich in Organic Matter:

This plant prefers soils saturated with plant organic matter. Its presence indicates a wet area due to surface water that cannot infiltrate or groundwater close to the surface, resulting in an accumulation of organic matter.

Preference for Slightly Acidic to Neutral Soils:

Filipendula ulmaria prefers soils with a pH range from slightly acidic to neutral, specifically a pH range of 5.5 to 7.0. Its presence can indicate stable soil pH, often slightly acidic.

Preference for Sunny to Partially Shaded Environments:

Meadowsweet grows well in sunny to partially shaded environments. Its presence may indicate open or semi-open areas with good light exposure in cool climates.

FOENICULUM VULGARE (COMMON FENNEL)

WELL-DRAINED SOILS
POOR SOILS
PH OF 6.0 TO ABOVE 8.0
OPEN OR SEMI-OPEN AREAS
PERIODS OF DROUGHT

Preference for Well-drained Soils:

Common fennel thrives in well-drained, even dry soils.

Tolerance for Poor to Moderately Nutrient-rich Soils:

This plant can develop in very poor soils, particularly those lacking animal organic matter.

Preference for Neutral to Alkaline soils:

Foeniculum vulgare prefers soils with a pH range from neutral to alkaline, specifically a pH range of 6.0 to above 8.0.

Preference for Sunny Environments:

Common fennel grows well in environments with full sun or partial shade. Its presence may indicate open or semi-open areas.

Drought Tolerance:

Foeniculum vulgare is very drought-resistant. Its presence may indicate environments that can experience periods of drought or have low water availability, particularly in warm climates.

FRAGARIA VESCA (WILD STRAWBERRY)

WELL-DRAINED SOILS
SOILS RICH IN PLANT ORGANIC MATTER
SOILS DEFICIENT IN ANIMAL ORGANIC MATTER
PH OF 5.0 TO 7.0
SEMI-SHADED TO OPEN AREAS
MODERATELY DISTURBED HABITATS
TEMPERATE CLIMATE CONDITIONS

Preference for Well-drained to Dry soils:

Wild strawberry indicates well-drained to dry soils.

Tolerance for Soils Rich in Organic Matter:

This plant prefers soils rich in bases and plant organic matter. Its presence may indicate a deficiency in nitrogen or animal organic matter. It is often found in areas with a layer of plant debris that impedes water, oxygen, and nutrient circulation to plant roots.

Preference for Slightly Acidic to Neutral Soils:

Fragaria vesca prefers soils with a pH range from slightly acidic to neutral, specifically a pH range of 5.0 to 7.0. Its presence can indicate stable soil pH suitable for many forest and open environment plants.

Preference for Partially Shaded to Sunny Environments:

Wild strawberry grows well in partially shaded to sunny environments. Its presence may indicate semi-shaded to open areas with good light exposure in cool climates.

Tolerance to Light Disturbances:

Fragaria vesca can tolerate some disturbance and colonize lightly disturbed areas. Its presence may indicate habitats that have undergone moderate disturbances, often related to human activity or natural events.

FRAXINUS EXCELSIOR (COMMON ASH)

MODERATELY MOIST SOILS
RICH SOILS
PH OF 6.0 TO 8.0
OPEN OR SEMI-OPEN AREAS
STABLE ECOSYSTEMS

Preference for Well-drained to Moderately Moist Soils:

Common ash thrives in well-drained to moderately moist soils. Its presence often indicates soil that retains sufficient moisture without being constantly waterlogged.

Preference for Nutrient-rich Soils:

This species prefers soils rich in bases, organic matter, and nutrients. Its presence can indicate fertile soil, often found in deciduous forests, forest edges, and floodplains.

Preference for Neutral to Slightly Alkaline Soils:

Fraxinus excelsior prefers soils with a pH range from neutral to slightly alkaline, specifically a pH range of 6.0 to 8.0. Its presence can indicate stable soil pH suitable for many forest plants.

Preference for Sunny to Partially Shaded Environments:

Common ash grows well in sunny to partially shaded environments. Its presence may indicate open or semi-open areas with good light exposure in cool climates.

Stable Ecosystems and Stream Edges:

Fraxinus excelsior is often found in young and mature forests, forest edges, and along streams, indicating a relatively stable ecosystem often associated with floodplains or wet areas.

GALANTHUS NIVALIS (SNOWDROP)

WELL-DRAINED TO MOIST SOILS
DEEP SOILS
RICH SOILS
PH OF 6.5 TO 7.5
SHADED ENVIRONMENTS
STABLE ECOSYSTEMS

Preference for Well-drained to Moist Soils:
Snowdrop thrives in well-drained to moist soils. Its presence often indicates a deep soil that retains sufficient moisture while avoiding water saturation.

Tolerance for Soils Rich in Organic Matter:
This plant prefers base-rich soils rich in organic matter and nitrogen. Its presence can indicate fertile soil, often found in deciduous forests and woodlands.

Preference for Neutral to Slightly Alkaline Soils:
Galanthus nivalis prefers soils with a pH range from neutral to slightly alkaline, specifically a pH range of 6.5 to 7.5. Its presence can indicate a stable soil pH suitable for many forest plants.

Preference for Shaded to Partially Shaded Environments:
Snowdrop grows well in shaded to partially shaded environments. Its presence may indicate undergrowth or forest edges with dense vegetation cover in cool climates.

Indication of Mature Forests:
Galanthus nivalis is often found in mature, well-established forests, indicating a relatively stable and undisturbed ecosystem.

GALEOPSIS TETRAHIT (COMMON HEMP-NETTLE)

WELL-DRAINED SOILS
BASE-RICH SOILS
SOILS RICH IN PLANT ORGANIC MATTER
SOILS DEFICIENT IN ANIMAL ORGANIC MATTER
PH OF 5.5 TO 7.0
OPEN OR SEMI-OPEN AREAS
DISTURBED HABITATS

Preference for Well-drained to Moderately Moist Soils:
Common hemp-nettle thrives in well-drained to moderately moist soils. Its presence often indicates soil that retains sufficient moisture without being constantly waterlogged.

Tolerance for Soils Rich in Organic Matter and Nutrients:
This plant indicates soil rich in bases, overly rich in plant organic matter, and deficient in animal organic matter. It suggests an environment evolving towards forest.

Preference for Slightly Acidic to Neutral Soils:
Galeopsis tetrahit prefers soils with a pH range from slightly acidic to neutral, specifically a pH range of 5.5 to 7.0. Its presence can indicate stable soil pH suitable for many plants in disturbed and open environments.

Preference for Sunny to Partially Shaded Environments:
Common hemp-nettle grows well in sunny to partially shaded environments. Its presence may indicate open or semi-open areas with good light exposure in cool climates.

Tolerance to Disturbance:
Galeopsis tetrahit is highly tolerant of disturbance and can quickly colonize disturbed areas. Its presence may indicate habitats that have undergone natural or human disturbances, such as agricultural lands, logging areas, and roadsides.

GALIUM APARINE (CLEAVERS)

WELL-DRAINED TO MOIST SOILS
VERY RICH SOILS
PH OF 6.0 TO 8.0
SEMI-SHADED TO SHADED AREAS
DISTURBED HABITATS

Preference for Well-drained to Moist Soils:

Cleavers thrives in well-drained to moist soils. Its presence often indicates soil that retains sufficient moisture without being constantly waterlogged.

Tolerance for Soils Rich in Organic Matter and Nutrients:

This plant is a bio-indicator of soils very rich in organic matter and nutrients, especially nitrogen. Its presence can indicate fertile soil, often found in cultivated lands, vegetable gardens, gardens, roadsides, and fallows.

Preference for Slightly Alkaline Soils:

Galium aparine prefers soils with a pH range from neutral to slightly alkaline, specifically a pH range of 6.0 to 8.0. Its presence can indicate stable soil pH, often slightly basic.

Preference for Partially Shaded to Shaded Environments:

Cleavers grows well in partially shaded to shaded environments. Its presence may indicate semi-shaded to shaded areas with moderate to dense vegetation cover in cool climates.

Tolerance to Disturbance:

Galium aparine is highly tolerant of disturbance and can quickly colonize disturbed areas. Its presence may indicate habitats that have undergone natural or human disturbances, such as agricultural lands, gardens, and roadsides.

GENISTA TINCTORIA (DYER'S GREENWEED)

STRONG WATER CONTRASTS
POOR SOILS
PH OF 6 TO 8
OPEN AREAS
PERIODS OF DROUGHT

Preference for Well-drained to Dry Soils:

Dyer's greenweed thrives in well-drained to dry soils. Its presence often indicates soil with significant water contrasts, dry in summer and moist in winter.

Tolerance for Poor to Moderately Nutrient-rich Soils:

This plant can develop in soils poor in animal organic matter and nutrients, generally rich in plant organic matter, bases, and clay. Its presence can indicate under-grazed, low-fertility soil, often found in sandy areas, dry meadows, and heaths.

Preference for Neutral to Slightly Basic Soils:

Genista tinctoria prefers soils with a pH range from neutral to slightly basic, specifically a pH range of 6 to 8. Its presence can indicate stable soil pH suitable for many plants in open and disturbed environments.

Preference for Sunny Environments:

Dyer's greenweed grows well in environments with full sun exposure. Its presence may indicate open areas with full light exposure in warm climates.

Drought Tolerance:

Genista tinctoria is relatively drought-resistant. Its presence may indicate environments that can experience periods of drought or have limited water availability during certain times of the year.

GENTIANA LUTEA (YELLOW GENTIAN)

WELL-DRAINED SOILS
SOILS RICH IN BASES AND/OR LIMESTONE
PH OF 5.0 TO 7.0
OPEN OR SEMI-OPEN AREAS
MOUNTAINOUS CLIMATE

Preference for Well-drained to Moderately Moist Soils:

Yellow gentian thrives in well-drained to moderately moist soils. Its presence often indicates soil that retains sufficient moisture without being constantly waterlogged.

Tolerance for Soils Rich in Organic Matter:

This plant prefers soils rich in bases, rich in plant organic matter and moderately rich nutrients. It is often found in acidic mountain meadows and subalpine pastures.

Preference for Slightly Acidic to Slightly Basic Soils:

Gentiana lutea prefers soils with a pH range from slightly acidic to basic, specifically a pH range of 5.0 to 7.0. Its presence can indicate stable soil pH, often slightly acidic, but rich in bases or containing some limestone.

Preference for Sunny to Partially Shaded Environments:

Yellow gentian grows well in sunny to partially shaded environments. Its presence may indicate open or semi-open areas with good light exposure in mountainous climates.

GEUM URBANUM (WOOD AVENS)

WELL-DRAINED SOILS
RICH DEEP SOILS
PH OF 5.5 TO 7.0
SEMI-SHADED AREAS
LIGHTLY DISTURBED HABITATS

Preference for Well-drained to Moderately Moist Soils:

Wood avens thrives in well-drained to moderately moist soils. Its presence often indicates soil that retains sufficient moisture without being constantly waterlogged.

Tolerance for Soils Rich in Organic Matter:

This plant prefers deep soils rich in plant organic matter, humus, and nitrogen. Its presence can indicate soil evolving towards a forest state, typical of undergrowth, forest edges, and hedgerows.

Preference for Slightly Acidic to Slightly Alkaline Soils:

Geum urbanum prefers soils with a pH range from slightly acidic to alkaline, specifically a pH range of 5.5 to 7.0. Its presence can indicate a stable soil pH suitable for many forest plants.

Preference for Partially Shaded to Shaded Environments:

Wood avens grows well in partially shaded to shaded environments. Its presence may indicate semi-shaded areas with good vegetation cover in cool climates.

Tolerance to Light Disturbances:

Geum urbanum can tolerate light disturbances and colonize disturbed areas. It can grow on rubble and its presence may indicate habitats that have undergone moderate disturbances, often related to human activity or natural events.

GLECHOMA HEDERACEA (GROUND IVY)

MOIST SOILS
BASE-RICH SOILS
RICH SOILS
PH OF 5.0 TO 8.0
SEMI-OPEN AREAS
LIGHTLY DISTURBED HABITATS

Preference for Moist to Well-drained Soils:

Ground ivy thrives in moist to well-drained soils. Its presence often indicates soil that retains moisture without being constantly waterlogged.

Tolerance for Soils Rich in Organic Matter:

This plant favors base-rich soils saturated with plant organic matter and nitrogen. Its presence can indicate soil evolving towards a forest state.

Preference for Slightly Acidic to Slightly Basic Soils:

Glechoma hederacea prefers soils with a pH range from slightly acidic to slightly basic, specifically a pH range of 5.0 to 8.0. Its presence can indicate stable soil pH suitable for many plants in moist and forested environments.

Preference for Shaded to Partially Shaded Environments:

Ground ivy grows well in shaded to partially shaded environments. Its presence may indicate undergrowth, forest edges, or wet meadows with dense vegetation cover in cool climates.

Indication of Moderate to Low Disturbance:

This plant is often found in moderately to lightly disturbed habitats, such as roadsides, gardens, and wooded areas. It tolerates cutting very well and grows back from its creeping roots.

GLYCYRRHIZA GLABRA (LICORICE)

MOIST SOILS
RICH SOILS
PH OF 6.0 TO 8.0
OPEN AREAS
ATMOSPHERIC HUMIDITY

Preference for Moist Soils:

Licorice thrives in moist but well-drained soils. Its presence often indicates soil that avoids water saturation, allowing good aeration for the roots.

Tolerance for Nutrient-rich Soils:

This plant prefers soils rich in organic matter and nutrients. Its presence can indicate fertile soil, often found in meadows, riverbanks, and cultivated areas.

Preference for Neutral to Slightly Alkaline Soils:

Glycyrrhiza glabra prefers soils with a pH range from neutral to slightly alkaline, specifically a pH range of 6.0 to 8.0. Its presence can indicate stable soil pH, often slightly basic.

Preference for Sunny Environments:

Licorice grows well in environments with full sun exposure but with some atmospheric humidity in warm climates. Its presence can indicate open but moist areas with full light exposure.

HEDERA HELIX (COMMON IVY)

WELL-DRAINED SOILS
SOILS SATURATED WITH PLANT ORGANIC MATTER
SOILS DEFICIENT IN ANIMAL ORGANIC MATTER
PH OF 6.0 TO 7.5
PARTIALLY SHADED ENVIRONMENTS
RELATIVELY STABLE ECOSYSTEMS

Preference for Well-drained to Moist Soils:

Common ivy thrives in well-drained to moist soils. Its presence often indicates soil that retains sufficient moisture without being constantly waterlogged.

Tolerance for Soils rich in Organic Matter:

This plant favors soils saturated with plant organic matter and deficient in animal organic matter. Its presence indicates forest soil or soil evolving towards a forested state.

Preference for Slightly Acidic to Neutral Soils:

Hedera helix prefers soils with a pH range from slightly acidic to neutral, specifically a pH range of 6.0 to 7.5. Its presence can indicate stable soil pH suitable for many forest plants.

Preference for Shaded to Partially Shaded Environments:

Common ivy grows well in shaded to partially shaded environments in cool climates. Its presence may indicate undergrowth, forest edges, or shaded gardens with dense vegetation cover.

Indication of Stable and Undisturbed Environments:

Hedera helix is often found in stable and undisturbed habitats, indicating a relatively stable ecosystem, such as mature forests and established gardens.

HELLEBORUS NIGER (CHRISTMAS ROSE)

DRY SOILS
PRESENCE OF LIMESTONE
PH HIGHER THAN 7.0
PARTIALLY SHADED ENVIRONMENTS
LIGHTLY DISTURBED HABITATS

Preference for Dry Soils:

Christmas rose thrives in dry to arid soils. Its presence often indicates thin soil that cannot retain moisture.

Tolerance for Soils Rich in Organic Matter and Nutrients:

This plant prefers soils rich in bases and nutrients. It is often found in rocky undergrowth, forest edges, and mountains. It indicates the presence of limestone.

Preference for Neutral to Slightly Alkaline Soils:

Helleborus niger prefers calcareous soils with an alkaline pH, specifically higher than 7.0.

Preference for Partially Shaded to Shaded Environments:

Christmas rose grows well in partially shaded to shaded environments. Its presence may indicate semi-shaded to shaded areas with good vegetation cover in cool climates.

Tolerance to Light Disturbance:

Helleborus niger can tolerate light disturbances and colonize lightly disturbed areas. Its presence may indicate habitats that have undergone moderate disturbances, often related to human activity or natural events.

HERACLEUM SPHONDYLIUM (COMMON HOGWEED)

MOIST SOILS
RICH SOILS
FRESH DEEP SOILS
PH OF 6.0
OPEN OR SEMI-OPEN AREAS
TEMPERATE CLIMATE CONDITIONS
MODERATELY DISTURBED HABITATS

Preference for Moist Soils:

Common hogweed thrives in moist to moderately moist soils. Its presence often indicates soil that retains sufficient moisture without being constantly waterlogged.

Preference for Nutrient-rich Soils:

This plant prefers deep soils rich in bases, organic matter and nutrients. Its presence can indicate fertile soil, slightly saturated with animal or plant organic matter.

Preference for Neutral Soils:

Heracleum sphondylium prefers soils with a neutral pH, around 6.0. Its presence can indicate stable soil pH suitable for many meadow and forest plants.

Preference for Sunny to Partially Shaded Environments:

Common hogweed grows well in sunny to partially shaded environments. Its presence may indicate open or semi-open areas with good light exposure in cool climates.

Indication of Moderate Disturbance:

This plant is often found in disturbed habitats, such as meadows, roadsides, and fallows. Its presence can indicate habitat disturbance, often related to human activity.

HESPERIS MATRONALIS (DAME'S ROCKET)

MOIST DEEP SOILS
RICH SOILS
SOILS RICH IN PLANT ORGANIC MATTER
COMPACTED SOILS
PH OF 5.0 TO 7.0
SEMI-OPEN AREAS
DISTURBED HABITATS

Preference for Moist Soils:

Dame's rocket thrives in moist deep soils. Its presence often indicates a soil with a large water reserve without being constantly waterlogged.

Tolerance for Soils Rich in Organic Matter and Nutrients:

This plant prefers soils rich in bases, nutrients and fossilized plant organic matter. Its presence often indicates compacted soil.

Preference for Neutral to Slightly Acidic Soils:

Hesperis matronalis prefers soils with a pH range from neutral to slightly acidic, specifically a pH range of 5.0 to 7.0.

Preference for Sunny to Partially Shaded Environments:

Dame's rocket grows well in sunny to partially shaded environments. Its presence may indicate open or semi-open areas with good light exposure in cool climates.

Tolerance to Disturbance:

Hesperis matronalis is relatively tolerant of disturbance and can quickly colonize disturbed areas. Its presence may indicate habitats that have undergone natural or human disturbances, such as fallows, slopes, roadsides, forest edge and logging

HIERACIUM PILOSELLA (MOUSE-EAR HAWKWEED)

DRY SOILS
POOR SOILS
SOILS DEFICIENT IN ANIMAL ORGANIC MATTER
DEGRADED SOILS
PH OF 4.5 TO 8.0
SUNNY ENVIRONMENTS
PERIODS OF DROUGHT

Preference for Dry Soils:

Mouse-ear hawkweed thrives in well-drained to dry soils. Its presence often indicates sandy soil with low water retention and poor clay content.

Tolerance for Poor Soils:

This plant can develop in soils rich in bases but poor in nutrients and deficient in animal organic matter, sometimes overgrazed by sheep. Its presence can indicate arid and degraded soil.

Preference for Acidic to Neutral Soils:

Hieracium pilosella tolerates a wide range of soil pH, from acidic to basic, specifically a pH range of 4.5 to 8.0.

Preference for Sunny Environments:

Mouse-ear hawkweed grows well in environments with full sun exposure. Its presence indicates open areas.

Drought Tolerance:

Hieracium pilosella is relatively drought-resistant. Its presence can indicate environments that can experience periods of drought or have limited water availability.

HUMULUS LUPULUS (HOPS)

MOIST TO WATER-LOGGED SOILS
RICH SOILS
PH OF 6.0 TO 7.5
OPEN OR SEMI-OPEN AREAS
DISTURBED HABITATS

Preference for Moist to Water-logged Soils:

Hops thrive in moist soils. Its presence often indicates soil that retains much moisture without being constantly waterlogged.

Tolerance for Soils Rich in Organic Matter and Nutrients:

This plant prefers very rich soils, often saturated with organic matter and nutrients.

Preference for Neutral to Slightly Alkaline Soils:

Humulus lupulus prefers soils with a pH range from neutral to slightly alkaline, specifically a pH range of 6.0 to 7.5. Its presence can indicate stable soil pH, often slightly basic.

Preference for Sunny to Partially Shaded Environments:

Hops grow well in sunny to partially shaded environments. Its presence may indicate open or semi-open areas with good light exposure in cool climates.

Tolerance to Disturbance:

Humulus lupulus is relatively tolerant of disturbance and can quickly colonize disturbed areas. Its presence can indicate habitats that have undergone natural or human disturbances, such as riverbanks, moist forest edges and abandoned marshy agricultural lands.

HYACINTHOIDES NON-SCRIPTA (ENGLISH BLUEBELL)

MOIST DEEP SOILS
SOILS RICH IN PLANT ORGANIC MATTER
SOILS RICH IN HUMUS
PH OF 5.0 TO 7.0
SHADED ENVIRONMENTS
MATURE FOREST AND STABLE ECOSYSTEM

Preference for Moist to Well-drained Soils:

English bluebell thrives in deep and moist soils. Its presence often indicates soil that retains much moisture while allowing good root aeration.

Tolerance for Soils Rich in Organic Matter:

This plant prefers soils rich in plant organic matter and humus. Its presence can indicate relatively fertile soil, often found in deciduous forests and woodlands.

Preference for Slightly Acidic to Neutral Soils:

Hyacinthoides non-scripta prefers soils with a pH range from slightly acidic to neutral, specifically a pH range of 5.0 to 7.0. Its presence can indicate stable soil pH suitable for many woodland plants.

Preference for Shaded to Partially Shaded Environments:

English bluebell grows well in shaded to partially shaded environments in cool climates. Its presence may indicate undergrowth, forest edges, or clearings with dense vegetation cover.

Indication of Mature Forests and Stable Zones:

Like many forest flowers, Hyacinthoides non-scripta is often found in mature, well-established forests, indicating a relatively stable and undisturbed ecosystem.

HYPERICUM PERFORATUM (HYPERICUM PERFORATUM)

WELL-DRAINED SOILS
SOILS DEFICIENT IN ANIMAL ORGANIC MATTER
EXCESS OF PLANT ORGANIC MATTER
PROGRESSION TOWARDS FOREST CONDITIONS
PH OF 5.0 TO 7.0
OPEN OR SEMI-OPEN AREAS
MODERATELY DISTURBED HABITATS

Preference for Well-drained to Dry Soils:

St. John's wort thrives in well-drained to dry soils. Its presence often indicates soil that avoids water saturation while retaining sufficient moisture.

Tolerance for Poor to Moderately Nutrient-rich Soils:

This plant can develop in soils rich in bases but poor in animal organic matter and moderately rich in nutrients. Its presence indicates an accumulation of plant organic matter and a progression towards forest conditions.

Preference for Neutral to Slightly Acidic Soils:

Hypericum perforatum prefers soils with a pH range from slightly acidic to neutral, specifically a pH range of 5.0 to 7.0. Its presence can indicate a stable soil pH suitable for many meadow and forest plants.

Preference for Sunny to Partially Shaded Environments:

St. John's wort grows well in sunny to partially shaded environments in cool climates. Its presence may indicate open or semi-open areas with good light exposure.

Indication of Moderate Disturbance:

This plant is often found in disturbed habitats, such as meadows, forest edges, roadsides, and wastelands. Its presence can indicate habitat disturbance, often related to human activity.

ILEX AQUIFOLIUM (COMMON HOLLY)

WELL-DRAINED TO MOIST SOILS
HIGH ATMOSPHERIC HUMIDITY
SOILS RICH IN PLANT ORGANIC MATTER
PH OF 5.5 TO 7.0
SHADED TO PARTIALLY SHADED ENVIRONMENTS
STABLE ECOSYSTEM

Preference for Well-drained to Moist Soils:

Common holly thrives in well-drained to moist soils. Its presence often indicates soil that retains sufficient moisture without being constantly saturated. It lives near watercourses in warm climates and it always indicates a good atmospheric humidity.

Tolerance for Soils Rich in Plant Organic Matter:

This plant prefers soils rich or even saturated with plant organic matter and lacking animal organic matter. Its presence indicate forest soil.

Preference for Slightly Acidic to Neutral Soils:

Ilex aquifolium prefers soils with a pH range from slightly acidic to neutral, specifically a pH range of 5.5 to 7.0. Its presence can indicate stable soil pH suitable for many undergrowth plants.

Preference for Shaded to Partially Shaded Environments:

Common holly grows well in shaded to partially shaded environments.
Its presence may indicate undergrowth, forest edges, or hedgerows with dense vegetation cover in cool climates.

Indication of Mature Forests and Stable Zones:

Ilex aquifolium grows slowly and is found mainly in mature, well-established forests, indicating a relatively stable and undisturbed ecosystem.

INULA HELENIUM (ELECAMPANE)

MOIST SOILS
RICH SOILS
PH OF 5.0 TO 7.0
OPEN OR SEMI-OPEN AREAS

Preference for Moist to Well-drained Soils:

Elecampane thrives in moist to well-drained soils. Its presence often indicates soil that retains moisture very well while still allowing root aeration.

Tolerance for Soils Rich in Organic Matter:

This plant prefers soils rich in bases, clay, and plant organic matter. Its presence can indicate fertile and highly nutritious soil, but losing oxygen, often found in wet meadows, along watercourses, and in ditches.

Preference for Neutral to Slightly Acidic Soils:

Inula helenium prefers soils with a pH range from slightly acidic to neutral, specifically a pH range of 5.0 to 7.0. Its presence can indicate stable soil pH suitable for many plants in meadows and wet environments.

Preference for Sunny to Partially Shaded Environments:

Elecampane grows well in sunny to partially shaded environments. Its presence may indicate open or semi-open areas with good light exposure in cool climates.

IRIS PSEUDACORUS (YELLOW IRIS)

WET TO WATERLOGGED SOILS
LOSS OF OXYGEN
SOILS RICH IN ORGANIC MATTER
PH RANGE OF 5.0 TO 8.0
PERIODIC FLOODING OR HIGH WATER TABLE
SPRING RESURGENCE

Preference for Wet to Waterlogged Soils:

The yellow iris thrives in wet to waterlogged soils. Its presence often indicates a soil that retains much moisture, commonly found in marshes, riverbanks, ditches, and flood-prone areas. The soil is prone to hydromorphism (excess of water causing a loss of oxygen).

Tolerance to Soils Rich in Organic Matter:

This plant prefers soils rich in organic matter and nutrients. It is often found in wetlands with an accumulation of organic material.

Preference for Slightly Acidic to Slightly Basic Soils:

Iris pseudacorus prefers soils with a pH ranging from slightly acidic to neutral, specifically a pH range of 5.0 to 8.0.

Preference for Sunny to Partially Shaded Environments:

The yellow iris grows well in sunny to partially shaded environments. Its presence can indicate open or semi-open areas with good light exposure in cool climates.

Tolerance to Periodic Flooding:

Iris pseudacorus is tolerant of periodic flooding, making it an indicative species of frequently flooded areas or those with a high water table. It can be a sign of spring resurgence.

JUNCUS EFFUSUS (SOFT RUSH)

SOILS RICH IN ORGANIC PLANT MATTER
FORMATION OF GLEY
PH RANGE OF 3.5 TO 6.5
PERIODIC FLOODING OR HIGH WATER TABLE
SPRING RESURGENCE.

Preference for Wet to Waterlogged Soils:
The soft rush thrives in wet to waterlogged soils. Its presence often indicates soil that retains a lot of moisture, commonly found in marshes, riverbanks, ditches, and wet meadows. The soil is prone to hydromorphism.

Tolerance to Poor to Moderately Rich Nutrient Soils:
This plant develops in soils rich in organic plant matter. Its presence can indicate the formation of gley (reduction of metal oxides) in the soil due to moisture. It is intolerant to limestone.

Preference for Acidic Soils:
Juncus effusus prefers acidic soils, specifically a pH range of 3.5 to 6.5.

Preference for Sunny to Partially Shaded Environments:
The soft rush grows well in sunny to partially shaded environments. Its presence can indicate open or semi-open areas with good light exposure in cool climates.

Tolerance to Periodic Flooding:
Juncus effusus is tolerant of periodic flooding, making it an indicative species of frequently flooded areas or those with a high water table. It can be a good indicator of spring resurgence.

JUNIPERUS COMMUNIS (COMMON JUNIPER)

FILTERING TO DRY SOILS
SOILS RICH IN BASES
SOILS RICH IN PLANT ORGANIC MATTER
SOILS POOR IN ANIMAL ORGANIC MATTER
PH RANGE OF 4.5 TO 8.0
COOL TO COLD CLIMATE
PERIODS OF DROUGHT

Preference for Well-Drained Soils:

The common juniper thrives in well-drained to dry soils. Its presence often indicates soil with a light texture or very permeable soil experiencing intense droughts in summer and and intense cold in winter.

Tolerance to Poor and Thin to Moderately Rich Nutrient Soils:

This plant can develop in soils rich in bases, rich in plant organic matter but very poor in animal organic matter. Its presence can indicate very thin soil, often found in heaths, meadows, and rocky slopes.

Tolerance to a Wide Range of Soil pH:

Juniperus communis can develop in soils ranging from acidic to basic (with a pH of 4.5 to 8.0) but prefers alkaline soils with a pH above 7.5.

Preference for Sunny to Partially Shaded Environments:

The common juniper grows well in sunny to partially shaded environments. Its presence can indicate open or semi-open areas with good light exposure in cool to cold climates.

LACTUCA SERRIOLA (PRICKLY LETTUCE)

DRY TO ARID SOILS
SOILS ENRICHED WITH ANIMAL ORGANIC MATTER.
EXCESS NITROGEN, PRESENCE OF NITRATES.
PH RANGE OF 6.0 TO 8.0.
PERIODS OF DROUGHT.
DISTURBED HABITAT.

Preference for Dry Soils:

The prickly lettuce thrives in dry to arid soils. Its presence often indicates soil that does not retain water and experiences periods of drought.

Tolerance to Poor to Moderately Rich Nutrient Soils:

This plant can develop in soils rich in bases and enriched with animal organic matter or nitrogen. Its presence may indicate the presence of nitrates.

Preference for Neutral to Slightly Alkaline Soils:

Lactuca serriola prefers soils with a pH range from neutral to slightly alkaline, specifically a pH range of 6.0 to 8.0.

Preference for Sunny Environments:

Prickly lettuce grows well in environments with full sun exposure. Its presence can indicate open areas with full light exposure in warm climates.

Indication of Disturbance:

This plant is often found in disturbed habitats, such as wastelands, roadsides, and abandoned fields. Its presence can indicate habitat disturbance, often related to human activity.

LAMIUM ALBUM (WHITE DEAD-NETTLE)

MOIST TO WELL-DRAINED SOILS.
RICH SOILS.
OVERGRAZING.
PH RANGE OF 4.5 TO 7.5.
MODERATE HABITAT DISTURBANCE.

Preference for Moist to Well-Drained Soils:

The white dead-nettle thrives in moist to well-drained soils. Its presence often indicates soil that retains enough moisture without being constantly waterlogged.

Tolerance to Soils Rich in Organic Matter:

This plant prefers soils rich in bases, organic matter, nitrogen, and nutrients. Its presence may indicate overgrazing in moist environments.

Preference for Slightly Acidic to Neutral Soils:

Lamium album prefers soils with a pH range from slightly acidic to neutral, specifically a pH range of 4.5 to 7.5. Its presence may indicate a stable pH soil suitable for many wetland and forest plants.

Preference for Shaded to Partially Shaded Environments:

The white dead-nettle grows well in partially shaded environments. Its presence can indicate understory areas, forest edges, or gardens with vegetative cover in cool climates.

Indication of Moderate to Low Disturbance:

This plant is often found in habitats with moderate to low disturbance, such as path edges, gardens, and wooded areas. Its presence may indicate moderate habitat disturbance, often related to human activity.

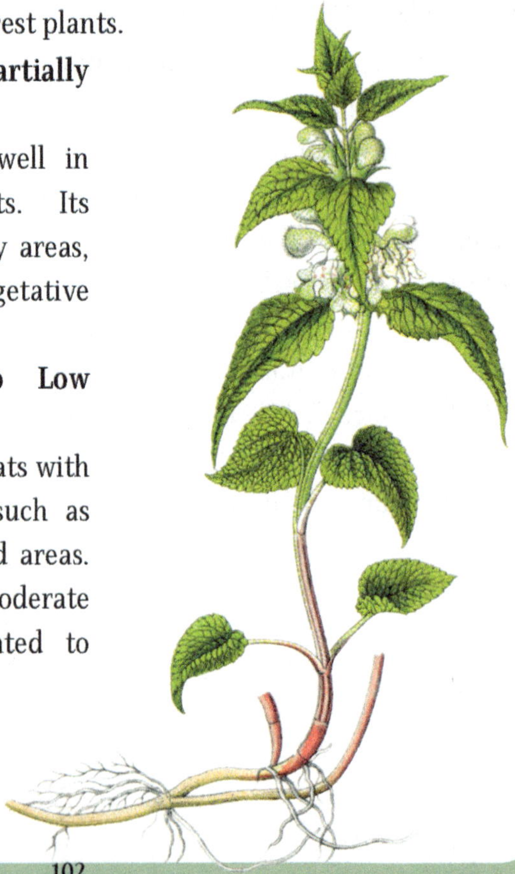

LATHYRUS PRATENSIS (MEADOW VETCHLING)

MOIST SOILS
SOILS RICH IN BASES
SOILS RICH IN ORGANIC MATTER
PH OF 5.5 TO 8.0
OPEN OR SEMI-OPEN AREAS
DISTURBED HABITAT.

Preference for Well-Drained to Moderately Moist Soils:

Meadow vetchling thrives in well-drained to moderately moist soils. Its presence indicates seasonal water contrasts and sometimes waterlogged soil.

Tolerance to Soils Rich in Organic Matter:

This plant prefers soils rich in bases and organic matter, as well as nutrients.

Preference for Neutral to Basic Soils:

Lathyrus pratensis prefers soils with a pH range from neutral to basic, specifically a pH range of 5.5 to 8.0. Its presence may indicate soil rich in bases or with a basic pH.

Preference for Sunny to Partially Shaded Environments:

Meadow vetchling grows well in sunny to partially shaded environments. Its presence can indicate open or semi-open areas with good light exposure in cool climates.

Tolerance to Disturbance:

Lathyrus pratensis is relatively tolerant to disturbance and can quickly colonize disturbed areas. Its presence may indicate a habitat that has undergone natural or human disturbances, such as meadows and roadsides.

LAVANDULA ANGUSTIFOLIA (TRUE LAVENDER)

DRY TO ARID SOILS
POOR AND ROCKY SOILS
CALCAREOUS SOILS
PH RANGE OF 6.5 TO 8.0
OPEN AREAS
PERIODS OF DROUGHT

Preference for Well-Drained to Dry Soils:

True lavender thrives in well-drained dry to arid soils. Its presence often indicates a very thin soil that does not retain water well, allowing good root aeration and preventing water saturation.

Tolerance to Poor to Moderately Rich Nutrient Soils:

This plant develops in soils that are poor to moderately rich in nutrients. Its presence indicates the presence of limestone in a low-fertility soil, deficient in humus, often found in rocky and calcareous soils.

Preference for Neutral to Slightly Alkaline Soils:

Lavandula angustifolia prefers soils with a pH range from neutral to slightly alkaline, specifically a pH range of 6.5 to 8.0. Its presence may indicate a stable and often slightly basic soil pH.

Preference for Sunny Environments:

True lavender grows well in environments with full sun exposure. Its presence can indicate open areas with full light exposure in warm climates.

Drought Tolerance:

Lavandula angustifolia is relatively drought-resistant. Its presence can indicate areas that experience periods of drought or have low water availability.

LEONTODON AUTUMNALIS (AUTUMN HAWKBIT)

MODERATELY MOIST SOILS
RICH SOILS
COMPACTED SOILS
PH RANGE OF 4.5 TO 7.5
OPEN OR SEMI-OPEN AREAS
MODERATELY DISTURBED HABITATS

Preference for Well-Drained to Moderately Moist Soils:

Autumn hawkbit thrives in moderately moist soils. Its presence often indicates soil that retains enough moisture without being constantly waterlogged.

Tolerance to Poor to Moderately Rich Nutrient Soils:

This plant can develop in compacted moist soils rich in bases and organic matter. Its presence may indicate overgrazing.

Preference for Neutral to Slightly Acidic Soils:

Leontodon autumnalis prefers soils with a pH range from slightly acidic to neutral, specifically a pH range of 4.5 to 7.5. Its presence may indicate soil with a stable pH suitable for many meadow plants.

Preference for Sunny to Partially Shaded Environments:

Autumn hawkbit grows well in sunny to partially shaded environments. Its presence can indicate open or semi-open areas with good light exposure.

Indication of Moderate to Low Disturbance:

This plant is often found in habitats with moderate to low disturbance, such as meadows, lawns, and roadsides. Its presence may indicate moderate habitat disturbance, often related to human activity.

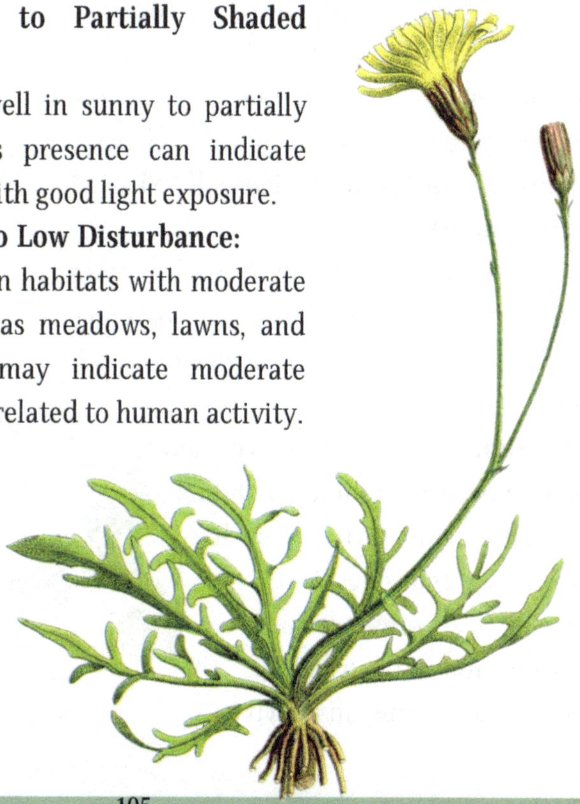

LEUCANTHEMUM VULGARE (OXEYE DAISY)

WELL-DRAINED SOILS.
SOILS RICH IN PLANT ORGANIC MATTER.
SOILS DEFICIENT IN ANIMAL ORGANIC MATTER.
PH RANGE OF 5.0 TO 7.0.
OPEN OR SEMI-OPEN AREAS.
MODERATELY DISTURBED HABITATS.

Preference for Well-Drained to Moderately Moist Soils:
The oxeye daisy thrives in well-drained to moderately moist soils. Its presence often indicates soil that retains sufficient moisture without being constantly waterlogged.

Tolerance to Moderately Rich Nutrient Soils:
This plant prefers soils rich in plant organic matter and deficient in animal organic matter. Its presence may indicate a medium-fertility soil, often found in meadows, roadsides, and lawns.

Preference for Neutral to Slightly Acidic Soils:
Leucanthemum vulgare prefers soils with a pH range from slightly acidic to neutral, specifically a pH range of 5.0 to 7.0. Its presence may indicate stable pH soil, suitable for many meadow and lawn plants.

Preference for Sunny to Partially Shaded Environments:
The oxeye daisy grows well in sunny to partially shaded environments. Its presence can indicate open or semi-open areas with good light exposure.

Indication of Moderate Disturbance:
This plant is often found in habitats with moderate disturbance, such as meadows, roadsides, and wastelands. Its presence may indicate habitat disturbance, often related to human activity.

LIGUSTRUM VULGARE (OXEYE DAISY)

WELL-DRAINED SOILS
MODERATELY RICH SOILS
PH OF 6.0 TO 8.5
OPEN OR SEMI-OPEN AREAS
DISTURBED HABITAT

Preference for Well-Drained to Moderately Moist Soils:

Common privet thrives in well-drained to moderately moist soils. Its presence often indicates soil that retains sufficient moisture without being constantly waterlogged.

Tolerance to Moderately Rich Nutrient Soils:

This plant prefers soils that are low in animal organic matter and moderately rich in nutrients.

Preference for Neutral to Alkaline Soils:

Ligustrum vulgare prefers soils with a pH range from neutral to alkaline, specifically a pH range of 6.0 to 8.5.

Preference for Sunny to Partially Shaded Environments:

Common privet grows well in sunny to partially shaded environments. Its presence can indicate open or semi-open areas with good light exposure in climates with warm summer.

Tolerance to Disturbance:

Ligustrum vulgare is relatively tolerant to disturbance and can quickly colonize disturbed areas. Its presence may indicate habitats that have undergone natural or human disturbances, such as hedgerows and roadsides.

LINARIA VULGARIS (COMMON TOADFLAX)

DRY SOILS
POOR SOILS
PH RANGE OF 6.0 TO 8.0.
SUNNY ENVIRONMENTS.
PERIODS OF DROUGHT.
DISTURBED HABITAT.

Preference for Well-Drained to Dry Soils:

Common toadflax thrives in well-drained to dry soils. Its presence often indicates rocky or sandy soil that does not retain water well.

Tolerance to Poor to Moderately Rich Nutrient Soils:

This plant can develop in poor thin soils that are deficient in organic matter and moderately rich in nutrients. Its presence indicates low-fertility soil, prone to erosion, loss of humus, or compaction.

Preference for Neutral to Slightly Alkaline Soils:

Linaria vulgaris prefers soils with a pH range from neutral to slightly alkaline, specifically a pH range of 6.0 to 8.0. Its presence may indicate a stable, often slightly basic pH soil.

Preference for Sunny Environments:

Common toadflax grows well in environments with full sun exposure. Its presence can indicate open areas with full light exposure in warm climates.

Drought Tolerance:

Linaria vulgaris is relatively drought-resistant. Its presence can indicate areas that experience periods of drought or have low water availability.

Indication of Disturbance:

This plant is often found in disturbed habitats, such as wastelands, roadsides, and abandoned fields. Its presence may indicate habitat disturbance, often related to human activity.

LINNAEA BOREALIS (TWINFLOWER)

WELL-DRAINED SOILS
NUTRIENT-POOR SOILS
VERY ACIDIC SOILS: PH OF 4.0 TO 5.5
SEMI-SHADED TO SHADED AREAS
COLD CLIMATIC CONDITIONS

Preference for Well-Drained to Moderately Moist Soils:

Twinflower thrives in well-drained to moderately moist soils. Its presence often indicates soil that retains sufficient moisture without being constantly waterlogged.

Tolerance to Acidic and Nutrient-Poor Soils:

This plant prefers very acidic and nutrient-poor soils. Its presence may indicate relatively infertile soil, suitable for few plants and often found in coniferous forests.

Preference for Slightly Acidic to Acidic Soils:

Linnaea borealis prefers soils with a pH range from moderately to strongly acidic, specifically a pH range of 4.0 to 5.5.

Preference for Partially Shaded to Shaded Environments:

Twinflower grows well in partially shaded to shaded environments. Its presence can indicate semi-shaded to shaded areas with good forest cover in cold climates.

LONICERA PERICLYMENUM (WOODBINE)

WELL-DRAINED SOILS
SOILS RICH IN PLANT ORGANIC MATTER
PH OF 5.0 TO 7.0
SHADED TO PARTIALLY SHADED ENVIRONMENTS
RELATIVELY STABLE ECOSYSTEM.

Preference for Well-Drained to Moderately Moist Soils:

Woodbine thrives in well-drained to moderately moist soils. Its presence often indicates soil that retains sufficient moisture without being constantly waterlogged.

Tolerance to Soils Rich in Plant Organic Matter:

This plant prefers base-poor soils rich in plant organic matter. It is often found in deciduous forests, hedgerows, and forest edges.

Preference for Slightly Acidic to Neutral Soils:

Lonicera periclymenum prefers soils with a pH range from slightly acidic to neutral, specifically a pH range of 5.0 to 7.0. Its presence may indicate stable pH soil suitable for many forest plants.

Preference for Partially Shaded Environments:

Woodbine grows well in shaded to partially shaded environments. Its presence can indicate understory areas, forest edges, or hedgerows with dense vegetative cover in cool climates.

Indication of Stable and Undisturbed Habitats:

Lonicera periclymenum is often found in stable and undisturbed habitats, which may indicate a relatively stable ecosystem, such as mature forests and established hedgerows.

LOTUS CORNICULATUS (BIRDSFOOT TREFOIL)

WELL-DRAINED SOILS
SOILS RICH IN BASES
BALANCED SOILS
PHOSPHORUS BLOCKAGE
PH RANGE OF 6.0 TO 8.0
PERIODS OF DROUGHT

Preference for Well-Drained to Moderately Moist Soils:

Birdsfoot trefoil thrives in well-drained to moderately moist soils. Its presence often indicates soil that retains sufficient moisture and is often waterlogged in winter.

Tolerance to Poor to Moderately Rich Nutrient Soils:

This plant can develop in soils rich in bases and balanced in organic matter and nutrients. Its presence may indicate stable and harmonious soil conditions with potential phosphorus blockage.

Preference for Neutral to Slightly Alkaline Soils:

Lotus corniculatus prefers soils with a pH range from neutral to slightly alkaline, specifically a pH range of 6.0 to 8.0. Its presence may indicate stable pH soil.

Preference for Sunny to Partially Shaded Environments:

Birdsfoot trefoil grows well in sunny to partially shaded environments. Its presence can indicate open or semi-open areas with moderate light exposure.

Drought Tolerance:

Lotus corniculatus is relatively drought-resistant. Its presence can indicate areas that experience periods of drought or have low water availability.

LUZULA SYLVATICA (GREATER WOOD-RUSH)

WELL-DRAINED SOILS
SOILS RICH IN PLANT ORGANIC MATTER
BASE-POOR SOILS
PH RANGE OF 5.0 TO 7.0
SHADED ENVIRONMENTS

Preference for Well-Drained to Moist Soils:

Greater wood-rush thrives in well-drained to moist soils. Its presence often indicates soil that retains sufficient moisture without being constantly waterlogged.

Tolerance to Soils Rich in Organic Matter:

This plant prefers soils rich in plant organic matter and poor in bases. It is often found in undergrowth and forest edges.

Preference for Slightly Acidic to Neutral Soils:

Luzula sylvatica prefers soils with a pH range from slightly acidic to neutral, specifically a pH range of 5.0 to 7.0. Its presence may indicate stable, often slightly acidic soil. This plant can sometimes be found in basic soils.

Preference for Shaded to Partially Shaded Environments:

Greater wood-rush grows well in shaded to partially shaded environments. Its presence can indicate semi-shaded to shaded areas with good vegetative cover in cool climates.

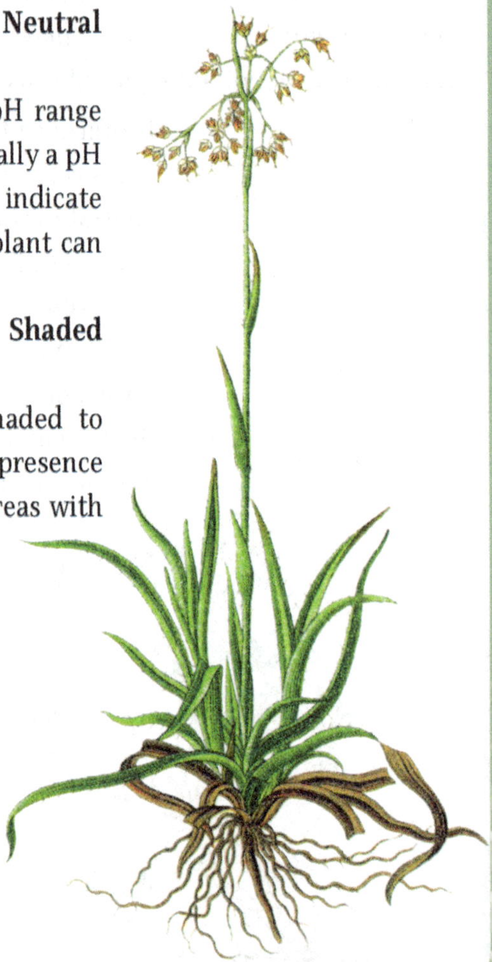

LYCOPUS EUROPAEUS (GYPSYWORT)

WATERLOGGED SOILS
SOILS RICH IN ORGANIC MATTER
PH RANGE OF 5.5 TO 7.0
CONSTANT OR PERIODIC PRESENCE OF WATER

Preference for Wet to Waterlogged Soils:

Gypsywort thrives in wet to waterlogged soils. Its presence often indicates soil with high water saturation, aquatic or semi-aquatic environments such as marshes, riverbanks, and wetlands, suggesting constant or periodic water presence in the habitat.

Tolerance to Soils Rich in Organic Matter:

This plant prefers eutrophic soils rich in organic matter and nitrites. Its presence may indicate hydromorphic soils, often found in wet environments with an accumulation of organic matter and gley formation.

Preference for Slightly Acidic to Slightly Basic Soils:

Lycopus europaeus prefers soils with a pH range from slightly acidic to slightly basic, specifically a pH range of 5.5 to 7.0.

Preference for Partially Shaded to Sunny Environments:

Gypsywort grows well in partially shaded to sunny environments. Its presence can indicate open or semi-open areas with good light exposure in cool climates.

LYSIMACHIA NUMMULARIA (MONEYWORT)

WET TO WATERLOGGED SOILS
ACCUMULATION OF ORGANIC MATTER
PH RANGE OF 5.0 TO 8.0
OPEN OR SEMI-OPEN AREAS
DISTURBED HABITAT

Preference for Wet to Waterlogged Soils:
The presence of moneywort often indicates soil that retains a lot of moisture and can sometimes be waterlogged, commonly found in wet areas, riverbanks, ditches, and wet meadows. It is tolerant to abrupt variations in soil moisture.

Tolerance to Soils Rich in Organic Matter and Nutrients:
This plant prefers soils with an accumulation and saturation of organic matter, rich in nutrients. The soil is often peaty and rich in bases, often alkaline, or sometimes composed of clayey or clay-loam alluvium.

Preference for Neutral to Alkaline Soils:
Lysimachia nummularia prefers soils with a pH range from slightly acidic to slightly alkaline, specifically a pH range of 5.0 to 8.0.

Preference for Partially Shaded to Sunny Environments:
Moneywort grows well in partially shaded to sunny environments. Its presence can indicate open or semi-open areas with good light exposure in cool climates.

Tolerance to Disturbance:
Lysimachia nummularia is relatively tolerant to disturbance and can quickly colonize disturbed areas. Its presence may indicate a habitat that has undergone natural or human disturbances.

LYSIMACHIA VULGARIS (COMMON LOOSESTRIFE)

WET TO WATERLOGGED SOILS
SOILS RICH IN ORGANIC MATTER
PH RANGE OF 5.0 TO 8.0
OPEN OR SEMI-OPEN AREAS
DISTURBED HABITAT

Preference for Wet to Waterlogged Soils:

The common loosestrife thrives in wet to waterlogged soils. Its presence often indicates soil that retains a lot of moisture and hydromorphism, commonly found in wet areas, marshes, riverbanks, and ditches. It is tolerant to abrupt variations in soil moisture.

Tolerance to Soils Rich in Organic Matter and Nutrients:

This plant prefers soils with an accumulation and saturation of organic matter, rich in nutrients. It can sometimes suggest polluted water with an excessive nutrient content, particularly nitrogen. The soil is often peaty and rich in bases, often alkaline, or sometimes composed of clayey or clay-loam alluvium.

Preference for Acidic to Alkaline Soils:

Lysimachia vulgaris prefers soils with a pH range from acidic to alkaline, specifically a pH range of 5.0 to 8.0.

Preference for Sunny to Partially Shaded Environments:

The common loosestrife grows well in sunny to partially shaded environments. Its presence can indicate open or semi-open areas with good light exposure in cool climates.

Tolerance to Disturbance:

Lysimachia vulgaris is relatively tolerant to disturbance and can quickly colonize disturbed areas. Its presence may indicate a habitat that has undergone natural or human disturbances.

LYTHRUM SALICARIA (PURPLE LOOSESTRIFE)

WET TO WATERLOGGED SOILS
NUTRIENT-RICH SOILS
PH OF 4.5 TO 7.5
OPEN OR SEMI-OPEN AREAS
PERIODIC FLOODING OR HIGH WATER TABLE

Preference for Wet to Waterlogged Soils:

Purple loosestrife thrives in wet to waterlogged soils, with hydromorphism. Its presence often indicates soil that retains a lot of moisture, commonly found in marshes, riverbanks, ditches, and flood-prone areas.

Tolerance to Soils Rich in Organic Matter and Nutrients:

This plant prefers soils saturated with organic matter and nutrients, with gley formation. Its presence may indicate water pollution with an excessive nutrient content, particularly nitrogen and phosphate. The soil is often peaty, sometimes muddy, loamy, or sandy.

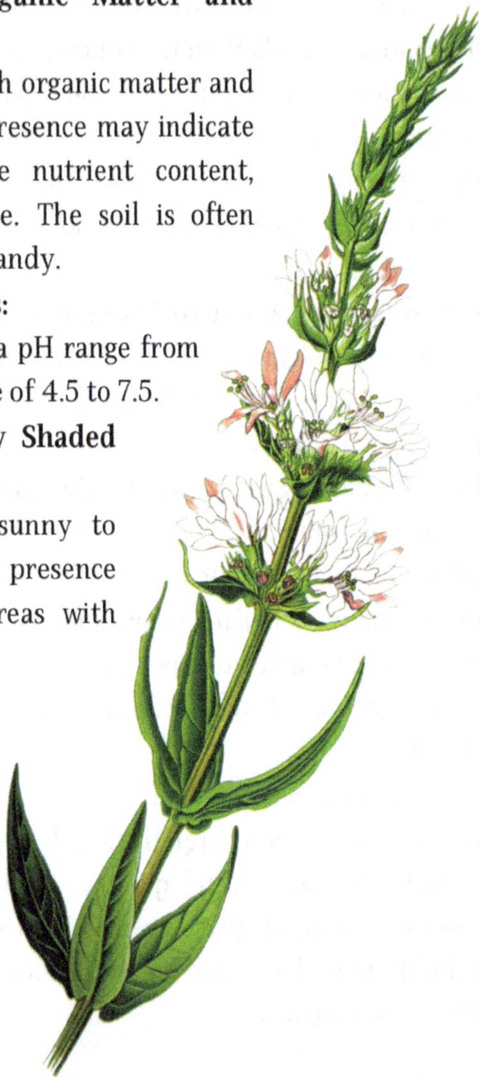

Preference for Acidic to Basic Soils:

Lythrum salicaria prefers soils with a pH range from acidic to basic, specifically a pH range of 4.5 to 7.5.

Preference for Sunny to Partially Shaded Environments:

Purple loosestrife grows well in sunny to partially shaded environments. Its presence can indicate open or semi-open areas with good light exposure in cool climates.

Tolerance to Periodic Flooding:

Lythrum salicaria is tolerant of periodic flooding, making it an indicative species of frequently flooded areas or those with a high water table.

MALUS SYLVESTRIS (WILD APPLE TREE)

WELL-DRAINED SOILS
SOILS DEFICIENT IN HUMUS
SOILS RICH IN BASES
PH OF 5.5 TO 8.5
OPEN OR SEMI-OPEN AREAS
STABLE AND UNDISTURBED HABITATS

Preference for Well-Drained to Moderately Moist Soils:

The wild apple tree thrives in well-drained to moderately moist soils. Its presence often indicates soil that does not retain water well but allows good root aeration.

Tolerance to Moderately Rich Nutrient Soils:

This plant prefers soils moderately rich in nutrients. Its presence may indicate soil deficient in humus, often found at forest edges, hedgerows, and clearings. The soil is often rich in bases.

Preference for Neutral to Slightly Acidic Soils:

Malus sylvestris prefers soils with a pH range from slightly acidic to alkaline, specifically a pH range of 5.5 to 8.5. Its presence may indicate stable pH soil, suitable for many forest and semi-open environment plants.

Preference for Sunny to Partially Shaded Environments:

The wild apple tree grows well in sunny to partially shaded environments. Its presence can indicate open or semi-open areas with good light exposure.

Indication of Stable and Undisturbed Habitats:

Malus sylvestris is often found in stable and undisturbed habitats, which may indicate a relatively stable ecosystem, such as forest edges and established hedgerows.

MALVA SYLVESTRIS (COMMON MALLOW)

DRY SOILS
SOILS POOR IN ORGANIC MATTER
SOILS RICH IN BASES
SOILS RICH IN NITROGEN AND POTASH
PH RANGE OF 6.0 TO 7.5
OPEN OR SEMI-OPEN AREAS
DISTURBED HABITATS

Preference for Well-Drained to Moderately Moist Soils:

Common mallow thrives in well-drained to dry soils.

Tolerance to Soils Poor in Organic Matter and Rich in Nutrients:

This plant prefers soils poor in organic matter but very rich in bases. The soil is also very rich, even saturated with nitrogen and potash.

Preference for Neutral to Slightly Alkaline Soils:

Malva sylvestris prefers soils with a pH range from neutral to slightly alkaline, specifically a pH range of 6.0 to 7.5. Its presence may indicate stable pH soil, often slightly basic.

Preference for Sunny to Partially Shaded Environments:

Common mallow grows well in sunny to partially shaded environments. Its presence can indicate open or semi-open areas with good light exposure in cool climates.

Tolerance to Disturbance:

Malva sylvestris is relatively tolerant to disturbance and can quickly colonize disturbed areas. Its presence may indicate a habitat that has undergone natural or human disturbances, such as wastelands, roadsides, and cultivated lands.

MATRICARIA CHAMOMILLA (WILD CHAMOMILE)

SOILS WITH SIGNIFICANT WATER CONTRASTS
SOILS POOR IN ORGANIC MATTER
SOILS RICH IN BASES
SOILS RICH IN NITROGEN
PH RANGE OF 5.5 TO 8.0
SUNNY ENVIRONMENTS
PERIODS OF LOW WATER AVAILABILITY
DISTURBED HABITATS

Preference for Well-Drained Soils:

Wild chamomile thrives in well-drained to very moist soils. Its presence can indicate significant water contrasts, erosion, and leaching.

Tolerance to Soils Poor in Organic Matter but Rich in Nutrients:

This plant can develop in soils poor in organic matter but rich in nutrients. The soil is rich in bases and can be saturated with nitrogen. It tolerates a certain amount of salt and can indicate soil salinization.

Preference for Neutral to Slightly Alkaline Soils:

Matricaria chamomilla prefers soils with a pH range from neutral to slightly alkaline, specifically a pH range of 5.5 to 8.0. Its presence may indicate stable pH soil, often slightly basic.

Preference for Sunny Environments:

Wild chamomile grows well in environments with full sun exposure. Its presence can indicate open areas with full light exposure.

Drought Tolerance:

Matricaria chamomilla is relatively drought-resistant. Its presence can indicate areas that experience periods of drought or have low water availability at certain times.

Indication of Disturbance:

This plant is often found in disturbed habitats, such as wastelands, roadsides, and abandoned fields. Its presence can indicate habitat disturbance, often related to human activity.

MEDICAGO LUPULINA (BLACK MEDIC)

WELL-DRAINED TO DRY SOILS
SOILS RICH IN BASES
PHOSPHORUS BLOCKAGE DUE TO COMPACTION
DEFICIENCY IN ANIMAL ORGANIC MATTER
PH RANGE OF 6.0 TO 8.5
OPEN OR SEMI-OPEN AREAS
DISTURBED HABITATS

Preference for Well-Drained to Dry Soils:

Black medic thrives in well-drained to dry soils. Its presence often indicates soil that retains sufficient moisture but tends to be dry.

Tolerance to Moderately Rich Soils:

This plant prefers soils poor in organic matter but rich in bases. Its presence may indicate phosphorus blockage due to soil compaction or high pH, as well as a deficiency in animal organic matter.

Preference for Neutral to Alkaline Soils:

Medicago lupulina prefers soils with a pH range from neutral to alkaline, specifically a pH range of 6.0 to 8.5.

Preference for Sunny Environments:

Black medic grows well in sunny to partially shaded environments. Its presence can indicate open or semi-open areas with good light exposure.

Tolerance to Disturbance:

Medicago lupulina is relatively tolerant to disturbance and can quickly colonize disturbed areas. Its presence may indicate habitats that have undergone natural or human disturbances, such as agricultural lands, roadsides, and wastelands.

MELILOTUS OFFICINALIS (YELLOW SWEET CLOVER)

WELL-DRAINED SOILS
SOILS RICH IN BASES
COMPACTED AND ERODED SOILS
PH OF 6.5 TO 8.5
OPEN AREAS
PERIODS OF DROUGHT
DISTURBED HABITATS.

Preference for Well-Drained Soils:

Yellow sweet clover thrives in well-drained soils. Its presence often indicates soil that retains very little water.

Tolerance to Poor to Moderately Rich Nutrient Soils:

This plant can develop in soils rich in bases, subject to erosion and compaction. Its presence may indicate phosphorus blockage due to compaction or high pH.

Preference for Neutral to Slightly Alkaline Soils:

Melilotus officinalis prefers soils with a pH range from neutral to slightly alkaline, specifically a pH range of 6.5 to 8.5.

Preference for Sunny Environments:

Yellow sweet clover grows well in environments with full sun exposure. Its presence can indicate open areas with full light exposure.

Drought Tolerance:

Melilotus officinalis is relatively drought-resistant. Its presence can indicate areas that experience periods of drought or have limited water availability.

Tolerance to Disturbance:

Melilotus officinalis is tolerant to disturbance and can quickly colonize disturbed areas. Its presence may indicate habitats that have undergone natural or human disturbances, such as abandoned agricultural lands and roadsides.

MELISSA OFFICINALIS (LEMON BALM)

INDICATORS OF ANCIENT HABITATS OR RUINS
WELL-DRAINED SOILS
RICH SOILS
PH RANGE OF 6.0 TO 8.5
OPEN OR SEMI-OPEN AREAS
LIGHTLY DISTURBED HABITATS

Preference for Well-Drained to Moderately Moist Soils:

Lemon balm thrives in well-drained to moderately moist soils. Its presence often indicates soil that retains sufficient moisture while allowing good root aeration.

Indicator of Human Habitation:

This plant prefers areas originally inhabited by humans, making it a good indicator of ancient ruins or old habitats. It thrives in soils rich in bases and nutrients.

Preference for Neutral to Alkaline Soils:

Melissa officinalis prefers soils with a pH range from neutral to alkaline, specifically a pH range of 6.0 to 8.5.

Preference for Sunny to Partially Shaded Environments:

Lemon balm grows well in sunny to partially shaded environments. Its presence can indicate open or semi-open areas with good light exposure.

Indication of Moderate to Low Disturbance:

This plant is often found in moderately to lightly disturbed habitats, such as gardens, orchards, and areas near human habitation. Its presence may indicate moderate habitat disturbance, often related to human activity.

MENTHA AQUATICA (WATER MINT)

WATERLOGGED SOILS
AQUATIC OR SEMI-AQUATIC ENVIRONMENTS
LACK OF OXYGEN AND HYDROMORPHISM
ACCUMULATION OF ORGANIC MATTER
PH OF 5.5 TO 7.0
OPEN OR SEMI-OPEN AREAS

Preference for Very Wet to Waterlogged Soils:

Water mint thrives in soils constantly saturated with water. Its presence often indicates soil with low oxygen levels and hydromorphism. Mentha aquatica is commonly found in aquatic or semi-aquatic environments, suggesting a constant or periodic presence of water in the habitat.

Tolerance to Soils Rich in Organic Matter:

This plant prefers soils very rich in organic matter and nutrients. Its presence may indicate soil with nitrates.

Preference for Slightly Acidic to Neutral Soils:

Mentha aquatica prefers soils with a pH range from slightly acidic to neutral, specifically a pH range of 5.5 to 7.0. Its presence may indicate stable pH soil suitable for many wetland plants.

Preference for Sunny to Partially Shaded Environments:

Water mint grows well in sunny to partially shaded environments. Its presence can indicate open or semi-open areas with good light exposure in cool climates.

MENTHA SPICATA (SPEARMINT)

MOIST SOILS.
RICH SOILS.
PH RANGE OF 6.0 TO 7.5.
OPEN OR SEMI-OPEN AREAS.
MODERATELY TO LIGHTLY DISTURBED HABITATS.

Preference for Well-Drained to Moderately Moist Soils:

Spearmint thrives in well-drained to moist soils. Its presence indicates the absence of drought periods and often suggests waterlogged soil with hydromorphisms.

Tolerance to Soils Rich in Organic Matter:

This plant prefers soils very rich in organic matter and nutrients. It can be an indicator of soil saturated with plant organic matter, deficient in animal organic matter, and rich in bases.

Preference for Neutral to Slightly Acidic Soils:

Mentha spicata prefers soils with a pH range from slightly acidic to neutral, specifically a pH range of 6.0 to 7.5. Its presence may indicate stable pH soil suitable for many wetland and cultivated plants.

Preference for Sunny to Partially Shaded Environments:

Spearmint grows well in sunny to partially shaded environments. Its presence can indicate open or semi-open areas with good light exposure.

Indication of Moderate to Low Disturbance:

This plant is often found in moderately to lightly disturbed habitats, such as gardens, wet meadows, and stream banks. Its presence may indicate moderate habitat disturbance, often related to human activity.

MYOSOTIS ARVENSIS (FIELD FORGET-ME-NOT)

RATHER DRY SOILS
POOR SOILS
LEACHED SOILS
PH OF 5.5 TO 7.0
OPEN OR SEMI-OPEN AREAS
MODERATELY TO LIGHTLY DISTURBED HABITATS

Preference for Well-Drained to Moderately Moist Soils:
Field forget-me-not thrives in well-drained soils. Its presence often indicates soil that does not retain moisture well and is subject to erosion.

Tolerance to Poor Soils:
This plant can develop in soils that are poor to moderately rich in nutrients. Its presence may indicate bare soil, deficient in organic matter and humus, and may signal soils irrigated in hot weather, causing salt accumulation.

Preference for Neutral to Slightly Acidic Soils:
Myosotis arvensis prefers soils with a pH range from slightly acidic to neutral, specifically a pH range of 5.5 to 7.0. Its presence may indicate stable pH soil suitable for many meadow and open area plants.

Preference for Sunny to Partially Shaded Environments:
Field forget-me-not grows well in sunny to partially shaded environments. Its presence can indicate open or semi-open areas with good light exposure.

Indication of Moderate to Low Disturbance:
This plant is often found in habitats with moderate to low disturbance, such as meadows, roadsides, and wastelands. Its presence may indicate moderate habitat disturbance, often related to human activity.

NARCISSUS PSEUDONARCISSUS (WILD DAFFODIL)

WELL-DRAINED SOILS
SOILS RICH IN PLANT ORGANIC MATTER
SOILS DEFICIENT IN ANIMAL ORGANIC MATTER
PH RANGE OF 6.0 TO 7.5
OPEN OR SEMI-OPEN AREAS
STABLE HABITATS

Preference for Well-Drained to Moderately Moist Soils:

The wild daffodil thrives in well-drained to moderately moist soils. Its presence often indicates soil that retains sufficient moisture without being constantly waterlogged.

Tolerance to Soils Rich in Plant Organic Matter:

This plant prefers soils rich in plant organic matter and deficient in animal organic matter.

Preference for Slightly Basic to Neutral Soils:

Narcissus pseudonarcissus prefers soils with a pH range from neutral to slightly basic, specifically a pH range of 6.0 to 7.5. Its presence may indicate stable pH soil suitable for many plants in open and semi-open habitats.

Preference for Sunny to Partially Shaded Environments:

The wild daffodil grows well in sunny to partially shaded environments. Its presence can indicate open or semi-open areas with good light exposure in cool climates.

Indication of Natural or Semi-Natural Habitats:

This plant is often found in relatively stable habitats, such as natural meadows, light woods, and forest edges. Its presence may indicate a relatively stable and undisturbed ecosystem.

NEPETA CATARIA (CATNIP)

WELL-DRAINED TO DRY SOILS
SOILS POOR IN ORGANIC MATTER
SOILS RICH IN NUTRIENTS
PH RANGE OF 6.0 TO 8.5
OPEN OR SEMI-OPEN AREAS
PERIODS OF LOW WATER AVAILABILITY
DISTURBED HABITATS

Preference for Well-Drained to Dry Soils:

Catnip thrives in well-drained to moderately moist soils. Its presence often indicates soil that retains enough moisture without being constantly waterlogged.

Tolerance to Moderately Rich to Rich Nutrient Soils:

This plant can develop in soils poor in organic matter but rich in nutrients. Its presence may indicate soil that is moderately fertile to fertile, often found in wastelands, gardens, roadsides, and vacant lots.

Preference for Neutral to Alkaline Soils:

Nepeta cataria prefers soils with a pH range from neutral to slightly alkaline, specifically a pH range of 6.0 to 8.5. Its presence may indicate stable pH soil, often slightly basic.

Preference for Sunny to Partially Shaded Environments:

Catnip grows well in sunny to partially shaded environments. Its presence can indicate open or semi-open areas with good light exposure.

Drought Tolerance:

Nepeta cataria is relatively drought-resistant. Its presence can indicate areas that experience periods of drought or have limited water availability.

Tolerance to Disturbance:

Catnip is tolerant to disturbance and can quickly colonize disturbed areas. Its presence may indicate habitats that have undergone natural or human disturbances, such as wastelands and roadsides.

ORIGANUM VULGARE (OREGANO)

WELL-DRAINED TO DRY SOILS
POOR SOILS
PH RANGE OF 6.0 TO 8.0
OPEN OR SEMI-OPEN AREAS
PERIODS OF LOW WATER AVAILABILITY
WARM CLIMATE

Preference for Well-Drained Soils:

Oregano thrives in well-drained soils. Its presence often indicates dry and rocky soil, avoiding water saturation and allowing good root aeration.

Tolerance to Poor to Moderately Rich Nutrient Soils:

This plant can develop in soils very poor in animal organic matter but rich in bases, often calcareous. Its presence may indicate soil that is not particularly fertile, often found in wastelands, dry grasslands, and rocky slopes.

Preference for Neutral to Slightly Alkaline Soils:

Origanum vulgare prefers soils with a pH range from neutral to slightly alkaline, specifically a pH range of 6.0 to 8.0. Its presence may indicate stable pH soil, often slightly basic.

Preference for Sunny to Partially Shaded Environments:

Oregano grows well in sunny to partially shaded environments. Its presence can indicate open or semi-open areas with good light exposure in warm climates.

Drought Tolerance:

Origanum vulgare is drought-resistant. Its presence can indicate areas that experience periods of drought or have low water availability.

OSMUNDA REGALIS (ROYAL FERN)

WATERLOGGED SOILS
PH OF 4.0 TO 7.0
SOILS RICH IN PLANT ORGANIC MATTER
PARTIALLY SHADED TO SHADED ENVIRONMENTS

Preference for Wet to Waterlogged Soils:
The royal fern often indicates soil that retains a lot of moisture, sometimes waterlogged, and is commonly found near marshes, riverbanks, ditches, and flood-prone areas.

Tolerance to Acidic to Neutral Soils:
This plant prefers acidic to neutral soils, with a pH range of 4.0 to 7.0. Its presence may indicate soil with a stable and often slightly acidic pH.

Tolerance to Soils Rich in Organic Matter:
Osmunda regalis prefers soils with an accumulation of plant organic matter.

Preference for Partially Shaded to Shaded Environments:
The royal fern grows well in partially shaded to shaded environments.
Its presence can indicate semi-shaded to shaded areas with good vegetative cover in cool climates.

OXALIS ACETOSELLA (WOOD SORREL)

WELL-DRAINED COOL SOILS
NUTRIENT-POOR SOILS
SOIL ACIDIFICATION
PH RANGE OF 4.0 TO 6.5
SHADED ENVIRONMENTS

Preference for Well-Drained to Moderately Moist Soils:

Wood sorrel thrives in well-drained to moderately moist soils. Its presence often indicates high atmospheric humidity and soil that retains sufficient moisture without being waterlogged.

Tolerance to Relatively Poor Soils:

This plant tolerates relatively nutrient-poor soils.

Preference for Acidic to Slightly Acidic Soils:

Oxalis acetosella prefers soils with a pH range from moderately acidic to slightly acidic, specifically a pH range of 4.0 to 6.5. Its presence may indicate soil undergoing acidification, often due to the presence of conifers.

Preference for Shaded to Partially Shaded Environments:

Wood sorrel grows well in shaded to partially shaded environments. Its presence can indicate shaded to semi-shaded areas with good vegetative cover in cool climates.

PAPAVER RHOEAS (WOOD SORREL)

WELL-DRAINED TO DRY SOILS
MODERATELY RICH SOILS
PH RANGE OF 6.0 TO 7.5
RECENT ELEVATION IN PH
OPEN AREAS
DISTURBED HABITATS
OFTEN LINKED TO SHEEP GRAZING OR PLOWING

Preference for Well-Drained to Dry Soils:

The common poppy thrives in well-drained soils. Its presence often indicates soil that does not excessively retain water, allowing good root aeration.

Tolerance to Moderately Rich Nutrient Soils:

This plant can develop in soils that are moderately rich in nutrients. Its presence may indicate soil that is not particularly fertile, often found in wastelands, roadsides, vacant lots, poor meadows, and cultivated fields.

Sudden increase in pH:

Papaver rhoeas prefers soils with a pH range from neutral to slightly alkaline, specifically a pH range of 6.0 to 7.5. Its presence can indicate a sudden increase in pH, often appearing after grazing by sheep, whose droppings have an alkalizing effect on the soil.

Preference for Sunny Environments:

The common poppy grows well in environments with full sun exposure. Its presence can indicate open areas with full light exposure.

Tolerance to Disturbance:

Papaver rhoeas is relatively tolerant to disturbance and can quickly colonize disturbed areas. It germinates when the seed is not buried in the soil and therefore often after plowing. It presence may thus indicate habitats that have undergone natural or human disturbances, especially related to soil alkalinization and agriculture.

PARIS QUADRIFOLIA (HERB PARIS)

MOIST SOILS
SOILS RICH IN BASES
PH OF 5.5 TO 8.5
SHADED ENVIRONMENTS
STABLE ECOSYSTEM
ANCIENT FOREST

Preference for Moist Soils:

The herb Paris quadrifolia thrives in moist soils and cool, high-humidity forests.

Tolerance for Soils Rich in Organic Matter:

This plant prefers soils moderately rich in nutrients, bases, and nitrogen. Its presence can indicate fertile soil, often found in deciduous forests, particularly beech woods, undergrowth, and shaded areas.

Preference for Neutral to Slightly Basic Soils:

Paris quadrifolia prefers soils with a pH range from slightly neutral to basic, from 5.5 to 8.5. Its presence can indicate soil with a stable pH, suitable for many forest plants.

Preference for Shaded to Partially Shaded Environments:

The herb grows well in shaded to partially shaded environments. Its presence can indicate undergrowth, forest edges, or densely wooded areas with dense vegetation cover in a cool climate.

Indication of Mature Forests and Stable Areas:

Paris quadrifolia is often found in mature and well-established forests, indicating a relatively stable and undisturbed ecosystem.

PEDICULARIS SYLVATICA (WOOD BETONY)

MOIST SOILS
SOILS RICH IN PLANT ORGANIC MATTER
POOR SOILS
PH OF 3.0 TO 7.0
SUNNY ENVIRONMENTS
DISTURBED HABITATS

Preference for Moist to Waterlogged Soils:

Wood betony thrives in moist to waterlogged soils. Its presence often indicates soil that retains a lot of moisture, with significant variations throughout the year, commonly found in wet meadows, marshes, and stream edges.

Tolerance for Nutrient-Poor Soils:

This plant can develop in soils rich in plant organic matter, very poor to moderately rich in nutrients, and poor in bases. Its presence can indicate soil that is not very fertile, often found in wet meadows and bogs.

Preference for Acidic Soils:

Pedicularis sylvatica prefers soils with an acidic pH, within a range of 3.0 to 6.0.

Preference for Sunny Environments:

Wood betony grows well in sunny environments. Its presence can indicate open or semi-open areas with good light exposure in a cool climate.

Tolerance to Light Disturbances:

Pedicularis sylvatica can tolerate light disturbances. Its presence can indicate a habitat that has experienced moderate natural or human disturbances, often related to agricultural activity or natural events.

PETASITES HYBRIDUS (BUTTERBUR)

MOIST SOILS
RICH SOILS
PH OF 6.0 TO 8.5
OPEN OR SEMI-OPEN AREAS
CONSTANT OR PERIODIC PRESENCE OF WATER

Preference for Moist to Waterlogged Soils:

Butterbur thrives in moist to waterlogged soils. Its presence often indicates soil with high water saturation and hydromorphic conditions, commonly found along riverbanks, ditches, and wet areas.

Tolerance for Soils Rich in Organic Matter:

This plant prefers soils rich in bases and very rich, even waterlogged, in organic matter and nutrients. Its presence outside its natural state can indicate an excess of animal organic matter or overgrazing in wet conditions.

Preference for Neutral to Alkaline Soils:

Petasites hybridus prefers soils with a pH range from neutral to alkaline, between 6.0 and 8.5.

Preference for Partially Shaded to Sunny Environments:

Butterbur grows well in partially shaded to sunny environments. Its presence can indicate open or semi-open areas with some light exposure in a cool climate.

Indication of Wet and Riparian Environments:

Petasites hybridus is a social and riparian species, colonizing riverbanks in large numbers. Its presence indicates a constant or periodic presence of water in the habitat.

PICEA ABIES (NORWAY SPRUCE)

WELL-DRAINED TO MOIST SOILS
MODERATELY RICH SOILS
PH OF 4.5 TO 7.0
OPEN OR SEMI-OPEN AREAS
COOL TO COLD CLIMATE
STABLE ECOSYSTEM

Preference for Well-Drained to Moist Soils:

Norway spruce thrives in well-drained to moist soils. Its presence often indicates soil that retains sufficient moisture while allowing good root aeration. It grows best in areas with high atmospheric humidity and is very sensitive to drought.

Tolerance for Soils Rich to Moderately Rich in Organic Matter:

This plant prefers soils that are rich to moderately rich in organic matter.

Preference for Acidic to Neutral Soils:

Picea abies prefers soils with a pH range from slightly acidic to neutral, between 4.5 and 7.0. The slow decomposition of its needles can acidify the soil and disrupt clay structure.

Preference for Partially Shaded to Sunny Environments:

Norway spruce grows well in partially shaded to sunny environments. It requires a lot of light as an adult but can tolerate shade in its youth. This tree thrives in cool to cold climates.

Indication of Mature Forests and Stable Areas:

Picea abies is often found in mature and well-established forests, indicating a relatively stable and undisturbed ecosystem.

PILOSELLA OFFICINARUM (MOUSE-EAR HAWKWEED)

DRY SOILS
POOR SOILS
PH OF 4.5 TO 7.0
SUNNY ENVIRONMENTS
PERIODS OF DROUGHT

Preference for Well-Drained to Dry Soils:

Mouse-ear hawkweed thrives in well-drained to dry soils. Its presence often indicates soil that does not retain water well, allowing good root aeration.

Tolerance for Poor to Moderately Nutrient-Rich Soils:

This plant can develop in soils that are poor to moderately rich in nutrients. Its presence can indicate soil that is not particularly fertile, often found in dry lawns, meadows, and sandy areas.

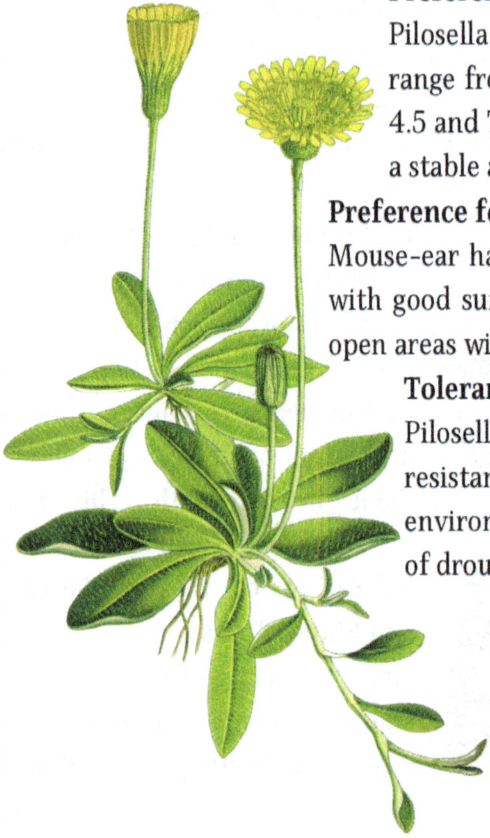

Preference for Acidic to Neutral Soils:

Pilosella officinarum prefers soils with a pH range from slightly acidic to neutral, between 4.5 and 7.0. Its presence can indicate soil with a stable and often slightly acidic pH.

Preference for Sunny Environments:

Mouse-ear hawkweed grows well in environments with good sun exposure. Its presence can indicate open areas with full light exposure.

Tolerance to Drought:

Pilosella officinarum is relatively drought-resistant. Its presence can indicate environments that may experience periods of drought or have limited water availability.

Tolerance to Light Disturbances:

Pilosella officinarum can tolerate light disturbances and colonize slightly disturbed areas. Its presence can indicate a habitat that has experienced moderate natural or human disturbances.

PINUS SYLVESTRIS (SCOTS PINE)

WELL-DRAINED SOILS
POOR SOILS
PH OF 4.0 TO 6.5
OPEN AREAS
SUMMER DROUGHTS

Preference for Well-Drained to Dry Soils:

Scots pine thrives in well-drained to dry soils. Its presence often indicates thin, rocky soils that are low in limestone, do not retain water well, allow good root aeration, and avoid water saturation. It tolerates but does not prefer waterlogged or compacted soils.

Tolerance for Poor to Moderately Nutrient-Rich Soils:

This plant can develop in poor to moderately nutrient-rich soils. Its presence can indicate soil that is not particularly fertile, often found in heaths, rocky slopes, and sandy soils.

Preference for Acidic to Neutral Soils:

Pinus sylvestris prefers soils with a pH range from slightly acidic to neutral, between 4.0 and 6.5. This tree does not thrive and declines on calcareous soil. The decomposition of its needles acidifies the soil.

Preference for Sunny Environments:

Scots pine grows well in environments with good sun exposure. Its presence can indicate open areas with full light exposure.

Tolerance to Drought:

Pinus sylvestris is relatively drought-resistant in cool climates during summer. Its presence can indicate environments that may experience periods of drought or have low water availability.

PLANTAGO LANCEOLATA (RIBWORT PLANTAIN)

MODERATELY MOIST SOILS
BALANCED AND HEALTHY SOILS
SOIL PH OF 5.5 TO 7.5
OPEN OR SEMI-OPEN AREAS

Preference for Well-Drained to Moderately Moist Soils:

Ribwort plantain thrives in well-drained to moderately moist soils. Its presence often indicates balanced soil that retains sufficient moisture without being constantly waterlogged.

Tolerance for Moderately Nutrient-Rich Soils:

This plant can develop in soils moderately rich in organic matter and nutrients. Its presence is a good sign, indicating healthy soil.

Preference for Neutral to Slightly Alkaline Soils:

Plantago lanceolata prefers soils with a pH range from slightly alkaline to neutral, between 5.5 and 7.5. Its presence can indicate soil with a stable pH, suitable for many plants in meadows and open environments.

Preference for Sunny to Partially Shaded Environments:

Ribwort plantain grows well in sunny to partially shaded environments. Its presence can indicate open or semi-open areas with good light exposure.

Indication of Moderate to Low Disturbance:

The presence of this plant can indicate moderate habitat disturbance, often related to human activity or grazing, and slightly compacted soils.

PLANTAGO MAJOR (BROADLEAF PLANTAIN)

ALL LEVELS OF SOIL MOISTURE
RICH SOILS
SOIL PH OF 5.0 TO 7.5
OPEN OR SEMI-OPEN AREAS
HIGHLY DISTURBED HABITATS
ANAEROBIC SOIL CONDITIONS
TRAMPLED SOIL

Tolerance for Soils from Dry to Waterlogged:

Broadleaf plantain thrives in a wide variety of soils, from dry to waterlogged.

Tolerance for Soils Rich in Organic Matter:

This plant prefers soils rich in organic matter, often resulting from grazing or human activity.

Preference for Neutral to Slightly Alkaline Soils:

Plantago major prefers soils with a pH range from slightly alkaline to neutral, between 5.0 and 7.5. Its presence can indicate soil with a stable pH, suitable for many plants in open and disturbed environments.

Preference for Sunny to Partially Shaded Environments:

Broadleaf plantain grows well in sunny to partially shaded environments. Its presence can indicate open or semi-open areas with good light exposure.

Indication of Moderate to High Disturbance:

The primary bio-indicator characteristic of this plant is its presence in disturbed and compacted habitats, such as lawns, trampled paths, roadsides, and urban areas. Its presence can indicate compacted and anaerobic soil due to trampling, overgrazing, or excessive soil working, especially in wet conditions.

POA PRATENSIS (KENTUCKY BLUEGRASS)

STRONG HYDRIC CONTRASTS
MODERATELY RICH SOILS
SOILS RICH IN BASES
SOIL PH OF 6.0 TO 8.0
OPEN OR SEMI-OPEN AREAS
DISTURBED HABITAT

Preference for Soils with Strong Hydric Contrasts:

Kentucky bluegrass thrives in well-drained to moderately moist soils. Its presence often indicates soil that retains sufficient moisture in winter but experiences significant drought in summer.

Tolerance for Moderately Rich to Rich Nutrient Soils:

This plant prefers soils that are moderately rich to rich in nutrients and bases. Its presence can indicate fertile soil, often found in meadows, pastures, lawns, and cultivated fields.

Preference for Neutral to Slightly Alkaline Soils:

Poa pratensis prefers soils with a pH range from neutral to slightly alkaline, between 6.0 and 8.0.

Preference for Sunny to Partially Shaded Environments:

Kentucky bluegrass grows well in sunny to partially shaded environments. Its presence can indicate open or semi-open areas with good light exposure in a cool climate.

Tolerance to Disturbance:

Poa pratensis is relatively tolerant to disturbance and can quickly colonize disturbed areas. Its presence can indicate habitats that have experienced natural or human disturbances, such as agricultural lands, lawns, and roadsides.

POLYGONATUM MULTIFLORUM (SOLOMON'S SEAL)

DRY SOILS
SOILS RICH IN BASES
SOILS SATURATED WITH PLANT ORGANIC MATTER
SOIL PH OF 5.5 TO 8.0
SHADED TO SEMI-SHADED AREAS

Preference for Well-Drained to Dry Soils:
Solomon's seal thrives in well-drained to dry soils. Its presence often indicates soil that does not retain water, which can be rocky or sandy, calcareous, or siliceous.

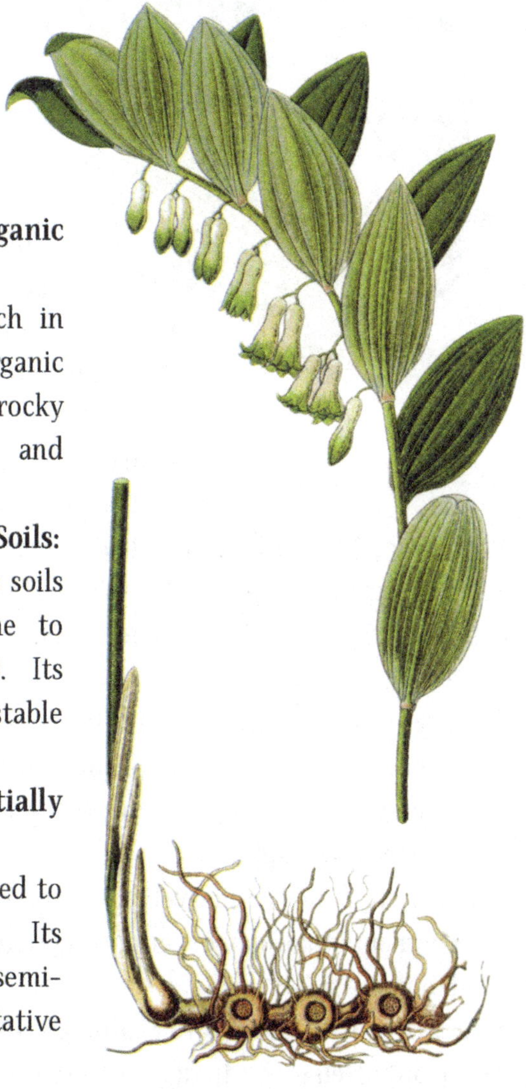

Tolerance for Soils Rich in Organic Matter:
This plant prefers soils very rich in bases and saturated with plant organic matter. It is often found in rocky undergrowth, forest edges, and hedgerows.

Preference for Neutral to Basic Soils:
Polygonatum multiflorum prefers soils with a pH range from alkaline to neutral, between 5.5 and 8.0. Its presence can indicate soil with a stable and often basic pH.

Preference for Shaded to Partially Shaded Environments:
Solomon's seal grows well in shaded to partially shaded environments. Its presence can indicate shaded to semi-shaded areas with good vegetative cover in a cool climate.

POPULUS TREMULA (ASPEN)

MOIST SOILS
SOIL PH OF 4.5 TO 7.0
SUNNY ENVIRONMENTS
DROUGHT TOLERANCE
NEWLY OPENED OR DISTURBED HABITATS

Preference for Well-Drained to Moderately Moist Soils:

Aspen thrives in well-drained to moderately moist soils. Its presence often indicates soil that retains sufficient moisture without being constantly waterlogged. It favors water and can indicate the presence of a water table or a spring.

Tolerance for Poor to Rich Nutrient Soils:

This plant can develop in soils ranging from poor to rich in nutrients. Aspen is an opportunistic species of neglected areas and is very flexible regarding soil fertility. It is often found in clearings, meadows, forest edges, and disturbed lands.

Preference for Acidic to Neutral Soils:

Populus tremula prefers soils with a pH range from slightly acidic to neutral, between 4.5 and 7.0. Its presence can indicate soil with a stable, often slightly acidic, pH.

Preference for Sunny to Partially Shaded Environments:

Aspen grows well in sunny to partially shaded environments. Its presence can indicate open or semi-open areas with good light exposure in a cool climate.

Tolerance to Drought:

Populus tremula is relatively drought-resistant. Its presence can indicate environments that may experience periods of drought or have variable water availability.

Indication of Disturbed to Stable Environments:

This plant often grows in disturbed to stable habitats, such as clearings, forest edges, meadows, and wastelands.

POTENTILLA ANSERINA (SILVERWEED)

MOIST SOILS
STRONG HYDRIC CONTRASTS
NUTRIENT-RICH SOILS
SOIL PH OF 5.5 TO 8.5
OPEN OR SEMI-OPEN AREAS
DISTURBED HABITATS

Preference for Moist Soils and Hydric Contrasts:

Silverweed thrives in moist to moderately moist soils. Its presence often indicates soil subjected to strong hydric contrasts, whether natural or induced by irrigation.

Tolerance for Nutrient-Rich Soils:

This plant can develop in soils rich in bases, organic matter, and nutrients. It often grows in areas frequented by birds.

Preference for Acidic to Basic Soils:

Potentilla anserina prefers soils with a pH range from slightly acidic to basic, between 5.5 and 8.5.

Preference for Sunny to Partially Shaded Environments:

Silverweed grows well in sunny to partially shaded environments. Its presence can indicate open or semi-open areas with good light exposure in a cool climate.

Tolerance to Disturbance:

Potentilla anserina is relatively tolerant to disturbance and can quickly colonize disturbed areas. Its presence can indicate habitats that have experienced natural or human disturbances, such as abandoned agricultural lands and roadsides.

POTENTILLA ERECTA (TORMENTIL)

MOIST SOILS
SOILS RICH IN PLANT ORGANIC MATTER
POOR NUTRIENT SOILS
SOIL PH OF 4.5 TO 6.0
OPEN OR SEMI-OPEN AREAS

Preference for Well-Drained to Moderately Moist Soils:
Tormentil thrives in well-drained to moderately moist soils. Its presence often indicates soil that retains sufficient moisture without being constantly waterlogged and experiences significant variations throughout the year.

Tolerance for Poor to Moderately Nutrient-Rich Soils:
This plant can develop in soils rich in plant organic matter and poor to moderately rich in nutrients and bases.

Preference for Acidic to Slightly Acidic Soils:
Potentilla erecta prefers soils with a pH range from moderately acidic to slightly acidic, between 4.5 and 6.0. Its presence can indicate soil with a stable and often acidic pH.

Preference for Sunny to Partially Shaded Environments:
Tormentil grows well in sunny to partially shaded environments. Its presence can indicate open or semi-open areas with good light exposure in a cool climate.

Tolerance to Light Disturbance:
Potentilla erecta can tolerate light disturbances and colonize slightly disturbed areas. Its presence can indicate a habitat that has experienced moderate natural or human disturbances.

PRIMULA VULGARIS (COMMON PRIMROSE)

MODERATELY MOIST SOIL
ABSENCE OF SEVERE DROUGHTS
SOILS RICH IN PLANT ORGANIC MATTER
NUTRIENT-RICH SOILS
SOIL PH OF 5.5 TO 7.5
SEMI-OPEN OR SHADED AREAS
STABLE HABITATS

Preference for Well-Drained to Moist Soils:

Common primrose thrives in well-drained to moist soils, showing some tolerance to moisture variations and prospering best in moderate conditions. It does not tolerate very severe droughts.

Tolerance for Soils Rich in Organic Matter:

This plant prefers soils rich or even saturated in plant organic matter. Its presence can indicate soil rich in bases and nutrients, often found in meadows, undergrowth, and shaded gardens.

Preference for Slightly Acidic to Slightly Basic Soils:

Primula vulgaris prefers soils with a pH range from slightly acidic to basic, between 5.5 and 7.5. Its presence can indicate soil with a stable pH, suitable for many plants in open and semi-open environments.

Preference for Partially Shaded to Shaded Environments:

Common primrose grows well in partially shaded to shaded environments. Its presence can indicate semi-open or shaded areas with dense vegetation cover in a cool climate.

Indication of Temperate Climatic Conditions:

Common primrose is common in temperate climatic zones, tolerating moderately cold winters and moderately warm summers. Its presence can indicate a temperate climate with distinct seasons.

Tolerance to Light Disturbance:

Potentilla erecta can tolerate light disturbances and colonize slightly disturbed areas. Its presence can indicate a habitat that has experienced moderate natural or human disturbances.

PRUNELLA VULGARIS (SELF-HEAL)

MODERATELY MOIST SOILS
SOILS RICH IN PLANT MATERIAL
NUTRIENT-RICH SOILS
SOIL PH OF 5.0 TO 7.0
OPEN OR SEMI-OPEN AREAS
DISTURBED HABITATS
PIONEER SPECIES

Preference for Moderately Moist Soils:

Self-heal thrives in well-drained to moderately moist soils. Its presence often indicates soil that retains sufficient moisture without being constantly waterlogged.

Tolerance for Nutrient-Rich Soils:

This plant can develop in nutrient-rich soils. Its presence can indicate soil rich in plant material, bases, and available nutrients. It is a pioneer species, often found in meadows, lawns, pathsides, clearings, logging areas, and riverbanks.

Preference for Neutral Soils:

Prunella vulgaris prefers soils with a pH range from slightly acidic to neutral, between 5.0 and 7.0. Its presence can indicate soil with a stable and often slightly acidic pH.

Preference for Sunny to Partially Shaded Environments:

Self-heal grows well in sunny to partially shaded environments. Its presence can indicate open or semi-open areas with good light exposure in a cool climate.

Tolerance to Disturbance:

Prunella vulgaris is relatively tolerant to disturbance and can quickly colonize disturbed areas. Its presence can indicate habitats that have experienced natural or human disturbances, such as meadows, lawns, and pathsides.

PRUNUS AVIUM (WILD CHERRY)

MODERATELY MOIST SOILS
SOILS RICH IN BASES
SOIL PH OF 6.0 TO 7.5
OPEN OR SEMI-OPEN AREAS
STABLE ECOSYSTEM

Preference for Well-Drained to Moderately Moist Soils:

Wild cherry thrives in moderately moist soils. Its presence often indicates soil that retains moisture well without becoming waterlogged.

Tolerance for Soils Rich in Organic Matter:

This plant prefers soils rich in bases and moderately rich in nutrients.

Preference for Neutral to Basic Soils:

Prunus avium prefers soils with a pH range from neutral to basic, between 6.0 and 7.5. Its presence can indicate soil with a stable, often slightly acidic, pH.

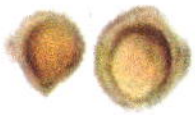

Preference for Sunny to Partially Shaded Environments:

Wild cherry grows well in sunny to partially shaded environments. Its presence can indicate open or semi-open areas with good light exposure in a cool climate.

Indication of Mature Forests and Stable Areas:

Prunus avium is often found in mature and well-established forests, indicating a relatively stable and undisturbed ecosystem.

PRUNUS SPINOSA (BLACKTHORN)

MODERATELY MOIST SOILS
SOILS VERY RICH IN BASES
POOR TO RICH SOILS
SOIL PH OF 6.0 TO 8.0
OPEN OR SEMI-OPEN AREAS
DISTURBED HABITATS
PIONEER SPECIES

Preference for Well-Drained to Moderately Moist Soils:
Blackthorn thrives in well-drained to moderately moist soils. Its presence often indicates soil that retains sufficient moisture without being constantly waterlogged.

Tolerance for Poor to Rich Nutrient Soils:
This plant prefers soils very rich in bases and can develop in soils ranging from poor to rich in nutrients. Its presence can indicate soil rich in plant organic matter, often found in hedgerows, forest edges, and meadows.

Preference for Neutral to Slightly Alkaline Soils:
Prunus spinosa prefers soils with a pH range from neutral to slightly alkaline, between 6.0 and 8.0. Its presence can indicate soil with a stable and often slightly basic pH.

Preference for Sunny to Partially Shaded Environments:
Blackthorn grows well in sunny to partially shaded environments. Its presence can indicate open or semi-open areas with good light exposure in a cool climate.

Tolerance to Disturbance:
Prunus spinosa is a pioneer species. It is tolerant to disturbance and can quickly colonize disturbed areas. Its presence can indicate habitats that have experienced natural or human disturbances, such as hedgerows, fallow lands, and roadsides.

PTERIDIUM AQUILINUM (BRACKEN FERN)

DRY TO MOIST SOILS
SOILS RICH IN PLANT ORGANIC MATTER
DEFICIENCY IN ANIMAL ORGANIC MATTER
SOIL PH OF 4.0 TO 6.5
OPEN OR SEMI-OPEN AREAS
PERIODS OF LOW WATER AVAILABILITY

Preference for Dry to Moist Soils:
Bracken fern thrives in dry to moist soils, showing significant flexibility regarding soil water availability.

Tolerance for Poor to Moderately Nutrient-Rich Soils:
This plant can develop in soils poor in nutrients and either poor or rich in bases. While the soil type is not critical, its presence often indicates an abundance of plant organic matter and a deficiency in nitrogen or animal organic matter.

Preference for Acidic to Neutral Soils:
Pteridium aquilinum prefers soils with a pH range from slightly acidic to neutral, between 4.0 and 6.5. It typically grows on acidic soils but can also be found on calcareous soils occasionally.

Preference for Sunny to Partially Shaded Environments:
Bracken fern grows well in sunny to partially shaded environments. Its presence can indicate open or semi-open areas with adequate light exposure in a cool climate.

Tolerance to Drought:
Pteridium aquilinum is relatively drought-resistant. Its presence can indicate environments that may experience periods of drought or have low water availability.

Indication of Moderate Disturbance:
This plant is often found in moderately disturbed habitats, such as heaths, meadows, logging areas, and light undergrowth. Its presence can indicate moderate habitat disturbance, often related to human activity or natural events.

PULMONARIA OFFICINALIS (COMMON LUNGWORT)

MODERATELY MOIST SOILS
RICH SOILS
SOIL PH OF 6.0 TO 8.0
SHADED TO SEMI-SHADED AREAS
LIGHTLY DISTURBED HABITATS
TEMPERATE CLIMATE

Preference for Well-Drained to Moderately Moist Soils:

Common lungwort thrives in well-drained to moderately moist soils. Its presence often indicates soil that retains sufficient moisture without being constantly waterlogged.

Tolerance for Soils Rich in Organic Matter:

This plant prefers soils rich in organic matter and nutrients. Its presence can indicate fertile soil, often found in undergrowth and forest edges.

Preference for Neutral to Slightly Alkaline Soils:

Pulmonaria officinalis prefers soils with a pH range from neutral to slightly alkaline, between 6.0 and 8.0. Its presence can indicate soil with a stable and often slightly basic pH.

Preference for Shaded to Partially Shaded Environments:

Common lungwort grows well in shaded to partially shaded environments. Its presence can indicate shaded to semi-shaded areas with good vegetative cover in a cool climate.

Tolerance to Light Disturbance:

Pulmonaria officinalis can tolerate light disturbances and colonize slightly disturbed areas. Its presence can indicate a habitat that has experienced moderate natural or human disturbances.

QUERCUS PETRAEA (SESSILE OAK)

WELL-DRAINED SOILS
SOIL PH OF 4.5 TO 6.5
OPEN OR SEMI-OPEN AREAS
STABLE ECOSYSTEM

Preference for Dry to Moderately Moist Soils:

Sessile oak thrives in dry to moderately moist soils. Its presence often indicates soil that retains sufficient moisture. While it tolerates dry soils well, it does not endure extreme droughts and can temporarily withstand waterlogged conditions.

Tolerance to Various Levels of Soil Fertility:

This tree prefers soils moderately rich in nutrients. It adapts to various levels of fertility and numerous soil types, often found in deciduous forests and forest edges.

Preference for Acidic to Neutral Soils:

Quercus petraea prefers soils with a pH range from slightly acidic to neutral, between 4.5 and 6.5. Its presence can indicate soil with a stable and often slightly acidic pH.

Preference for Sunny to Partially Shaded Environments:

Sessile oak grows well in sunny to partially shaded environments. Its presence can indicate open or semi-open areas with good light exposure in a cool climate.

Indication of Mature Forests and Stable Areas:

Quercus petraea is often found in mature and well-established forests, indicating a relatively stable and undisturbed ecosystem.

QUERCUS ROBUR (PEDUNCULATE OAK)

MODERATELY MOIST SOILS
SOILS RICH IN PLANT ORGANIC MATTER
DEFICIENCY IN ANIMAL ORGANIC MATTER
SOIL PH OF 5.0 TO 7.0
OPEN OR SEMI-OPEN AREAS
STABLE ECOSYSTEM

Preference for Well-Drained to Moderately Moist Soils:
Pedunculate oak thrives in well-drained to moderately moist soils. Its presence often indicates soil that retains sufficient moisture without being constantly waterlogged. This species does not tolerate severe droughts.

Tolerance for Rich to Moderately Nutrient-Rich Soils:
This plant prefers soils rich or even saturated with plant organic matter. Its presence can indicate a deficiency in animal organic matter, often found in deciduous forests, cool forests, and forest edges.

Preference for Neutral Soils:
Quercus robur prefers soils with a neutral pH, between 5.0 and 7.0. Its presence can indicate soil with a stable pH, suitable for many forest plants.

Preference for Sunny to Partially Shaded Environments:
Pedunculate oak grows well in sunny to partially shaded environments. Its presence can indicate open or semi-open areas with good light exposure in a cool climate.

Indication of Mature Forests and Stable Areas:
Quercus robur is often found in mature and well-established forests, indicating a relatively stable and undisturbed ecosystem.

RANUNCULUS ACRIS (MEADOW BUTTERCUP)

MOIST SOILS
NITROGEN-RICH SOILS
SOIL PH OF 5.0 TO 7.0
OPEN OR SEMI-OPEN AREAS

Preference for Well-Drained to Moist Soils:

Meadow buttercup thrives in well-drained to moist soils. Its presence often indicates soil that retains sufficient moisture without being constantly waterlogged. It does not tolerate drought well.

Tolerance for Soils Rich in Organic Matter:

This plant prefers soils fairly rich in bases and very rich in nitrogen. Its significant presence can indicate an excess of fertilizer or overgrazing, often found in cultivated areas, meadows, pastures, and roadsides.

Preference for Sunny to Partially Shaded Environments:

Meadow buttercup grows well in sunny to partially shaded environments. Its presence can indicate open or semi-open areas with good light exposure in a cool climate.

Preference for Neutral Soils:

Ranunculus acris prefers neutral soils with a pH range between 5.0 and 7.0. Its presence can indicate soil with a stable pH, suitable for many meadow and wetland plants.

Preference for Sunny to Partially Shaded Environments:

Meadow buttercup grows well in sunny to partially shaded environments. Its presence can indicate open or semi-open areas with good light exposure in a cool climate.

RANUNCULUS REPENS (RANUNCULUS REPENS)

WATERLOGGED SOILS
SOILS RICH IN ORGANIC MATTER
COMPACTED SOILS
HYDROMORPHIC CONDITIONS
SOIL PH OF 5.0 TO 7.0
OPEN OR SEMI-OPEN AREAS
DISTURBED HABITATS

Preference for Moist to Waterlogged Soils:

Creeping buttercup thrives in moist to waterlogged soils. Its presence often indicates soil that retains a lot of moisture, commonly found in wet meadows, ditches, riverbanks, and flood-prone areas.

Tolerance for Soils Rich in Organic Matter:

This plant prefers soils rich in organic matter and nutrients. Its presence indicates hydromorphic conditions and an accumulation of organic matter in compacted soil, often due to soil work or trampling in wet conditions.

Preference for Neutral Soils:

Ranunculus repens prefers neutral soils with a pH range between 5.0 and 7.0. Its presence can indicate soil with a stable and often slightly acidic pH.

Preference for Sunny to Partially Shaded Environments:

Creeping buttercup grows well in sunny to partially shaded environments. Its presence can indicate open or semi-open areas with good light exposure in a cool climate.

Tolerance to Disturbance:

Ranunculus repens is relatively tolerant to disturbance and can quickly colonize disturbed areas. Its presence can indicate habitats that have experienced natural or human disturbances, such as wet meadows, ditches, and riverbanks.

RHAMNUS CATHARTICA (COMMON BUCKTHORN)

WELL-DRAINED TO WATERLOGGED SOILS
SOILS RICH IN BASES
SOIL PH OF 6.0 TO 8.5
PRESENCE OF CALCAREOUS SOIL
OPEN OR SEMI-OPEN AREAS

Preference for Well-Drained to Moderately Moist Soils:

Common buckthorn thrives in well-drained to moderately moist soils. Its presence often indicates soil that retains sufficient moisture. It can also be found in very wet or marshy areas.

Tolerance for Soils Rich in Organic Matter:

This plant prefers soils rich in bases and moderately rich in organic matter and nutrients.

Preference for Neutral to Alkaline Soils:

Rhamnus cathartica prefers soils with a pH range from neutral to slightly alkaline, between 6.0 and 8.5. Its presence always indicates calcareous soil.

Preference for Sunny to Partially Shaded Environments:

Common buckthorn grows well in sunny to partially shaded environments. Its presence can indicate open or semi-open areas with good light exposure in a cool climate.

Indication of Moderate Disturbance:

This plant is often found in moderately disturbed habitats, such as fallow lands, hedgerows, and forest edges. Its presence can indicate moderate habitat disturbance, often related to human activity or natural events.

RHINANTHUS MINOR (YELLOW RATTLE)

DRY TO MODERATELY MOIST SOILS
POOR TO MODERATELY NUTRIENT-RICH SOILS
SOIL PH OF 5.0 TO 7.0
OPEN OR SEMI-OPEN AREAS
PLANT DIVERSITY

Preference for Dry to Moderately Moist Soils:
Yellow rattle thrives in dry to moderately moist soils. Its presence often indicates soil that retains sufficient moisture without being constantly waterlogged.

Tolerance for Poor to Moderately Nutrient-Rich Soils:
This plant can develop in soils that are poor to moderately rich in nutrients. Its presence can indicate soil that is not particularly fertile, often found in meadows, pastures, and roadsides.

Preference for Acidic to Neutral Soils:
Rhinanthus minor prefers soils with a pH range from slightly acidic to neutral, between 5.0 and 7.0.

Preference for Sunny to Partially Shaded Environments:
Yellow rattle grows well in sunny to partially shaded environments. Its presence can indicate open or semi-open areas with good light exposure in a cool climate.

Partial Parasitism:
Rhinanthus minor is a hemiparasitic plant that derives part of its nutrients from the roots of grasses and other herbaceous plants. Its presence can indicate sufficient plant diversity to support this growth strategy.

ROSA CANINA (DOG ROSE)

DRY TO MODERATELY MOIST SOILS
SOILS RICH IN PLANT ORGANIC MATTER
NITROGEN-DEFICIENT SOILS
SOILS RICH IN BASES
SOIL PH OF 6.0 TO 7.5
OPEN OR SEMI-OPEN AREAS

Preference for Dry to Moderately Moist Soils:

Dog rose thrives in dry to moderately moist soils but does not tolerate extreme droughts.

Tolerance for Rich to Moderately Nutrient-Rich Soils:

This plant prefers soils rich in bases and plant organic matter but deficient in nitrogen. It is often found in hedgerows, roadsides, and fallow lands.

Preference for Neutral to Slightly Alkaline Soils:

Rosa canina prefers soils with a pH range from neutral to slightly alkaline, between 6.0 and 7.5. Its presence can indicate soil with a stable and often slightly basic pH.

Rosa canina shows good tolerance to calcareous soils.

Preference for Sunny to Partially Shaded Environments:

Dog rose grows well in sunny to partially shaded environments. Its presence can indicate open or semi-open areas with good light exposure in a cool climate.

Indication of Disturbed to Semi-Natural Habitats:

This plant is often found in disturbed to semi-natural habitats, such as hedgerows, roadsides, and fallow lands. Its presence can indicate moderate habitat disturbance, often related to human activity.

ROSA RUBIGINOSA (SWEETBRIAR ROSE)

DRY SOILS
SOILS RICH IN BASES
SOILS RICH IN PLANT ORGANIC MATTER
NITROGEN-DEFICIENT SOILS
SOIL PH OF 6.0 TO 8.0
SUNNY ENVIRONMENTS
PERIODS OF LOW WATER AVAILABILITY

Preference for Dry Soils:
Sweet briar thrives in dry soils.

Tolerance for Poor to Rich Nutrient Soils:
This plant prefers soils rich in bases, saturated with plant organic matter, and deficient in nitrogen. It is often found in hedgerows, roadsides, meadows, and fallow lands.

Preference for Neutral to Slightly Alkaline Soils:
Rosa rubiginosa prefers soils with a pH range from neutral to slightly alkaline, between 6.0 and 8.0.

Preference for Sunny to Partially Shaded Environments:
Sweet briar grows well in sunny environments. Its presence can indicate open areas with good light exposure.

Tolerance to Drought:
Rosa rubiginosa is relatively drought-resistant. Its presence can indicate environments that may experience periods of drought or have limited water availability.

Tolerance to Disturbance:
Rosa rubiginosa is tolerant to disturbance and can quickly colonize disturbed areas. Its presence can indicate habitats that have experienced natural or human disturbances, such as fallow lands and roadsides.

ROSMARINUS OFFICINALIS (ROSEMARY)

DRY SOILS
SOILS RICH IN BASES
POOR SOILS
SOIL PH OF 6.0 TO 8.0
OPEN AREAS
DROUGHT CONDITIONS

Preference for Well-Drained to Dry Soils:

Rosemary thrives in rocky, well-drained to dry or very dry soils. Its presence often indicates very thin soil that does not retain water.

Tolerance for Poor to Moderately Nutrient-Rich Soils:

This plant can develop in soils that are poor to moderately rich in nutrients but very rich in bases. Its presence can indicate soil that is not particularly fertile, often found in rocky terrains, garrigues, and maquis.

Preference for Neutral to Alkaline Soils:

Rosmarinus officinalis prefers soils with a pH range from neutral to alkaline, between 6.0 and 8.0. Its presence can indicate soil with a stable and often slightly basic pH.

Preference for Sunny Environments:

Rosemary grows well in environments with full sun exposure. Its presence can indicate open areas with full light exposure in a warm climate.

Tolerance to Drought:

Rosmarinus officinalis is relatively drought-resistant. Its presence can indicate environments that may experience periods of drought or have low water availability.

RUBUS FRUTICOSUS (COMMON BRAMBLE)

MODERATELY MOIST SOILS
SOILS SATURATED IN PLANT ORGANIC MATTER
SOIL PH OF 5.0 TO 7.0
DISTURBED HABITATS

Preference for Well-Drained to Moderately Moist Soils:
Common bramble thrives in well-drained to moderately moist soils. Its presence often indicates soil that retains sufficient moisture without being constantly waterlogged.

Tolerance for Soils saturated with plant Organic Matter:
This plant prefers soils saturated with plant organic matter and moderately rich in nutrients. It indicates a soil in loss of nitrogen and more or less deficient in animal organic matter.

Preference for Neutral to Slightly Acidic Soils:
Rubus fruticosus prefers soils with a pH range from slightly acidic to neutral, between 5.0 and 7.0. Its presence can indicate soil with a stable pH, suitable for many plants in disturbed areas.

Preference for Sunny to Partially Shaded Environments:
Common bramble grows well in sunny to partially shaded environments. It fruits best in full sun, although it tolerates shade. Its presence can indicate open or semi-open areas with good light exposure.

Indication of Moderate to High Disturbance:
This plant is often found in disturbed habitats, such as hedgerows, roadsides, fallow lands, and abandoned areas. Its presence can indicate habitat disturbance, often related to human activity.

RUBUS IDAEUS (RASPBERRY)

MODERATELY MOIST SOILS
SOILS RICH IN ORGANIC MATTER AND NITROGEN
SOIL PH OF 5.5 TO 7.0
OPEN OR SEMI-OPEN AREAS
DISTURBED HABITATS

Preference for Well-Drained to Moderately Moist Soils:
Raspberry thrives in well-drained to moderately moist soils. Its presence often indicates soil that retains sufficient moisture without being constantly waterlogged.

Tolerance for Soils Rich in Organic Matter:
This plant prefers soils that are quite rich in bases, rich in plant organic matter, and relatively rich in nitrogen or animal organic matter. Its presence can indicate fertile soil, often found in undergrowth, forest edges, and gardens.

Preference for Slightly Acidic to Neutral Soils:
Rubus idaeus prefers soils with a pH range from slightly acidic to neutral, between 5.5 and 7.0. Its presence can indicate soil with a stable and often slightly acidic pH.

Preference for Sunny to Partially Shaded Environments:
Raspberry grows well in sunny to partially shaded environments. Its presence can indicate open or semi-open areas with good light exposure in a cool climate.

Tolerance to Disturbance:
Rubus idaeus is relatively tolerant to disturbance and can quickly colonize disturbed areas. Its presence can indicate habitats that have experienced natural or human disturbances, such as roadsides, clearings, and gardens.

RUMEX ACETOSA (COMMON SORREL)

MOIST SOILS
RICH AND BALANCED SOILS
SOIL PH OF 4.5 TO 7.0
OPEN OR SEMI-OPEN AREAS
RELATIVELY STABLE HABITATS

Preference for Moist to Well-Drained Soils:

Common sorrel thrives in moist to well-drained soils. Its presence often indicates soil that retains moisture well without being constantly waterlogged.

Preference for Soils Rich in Organic Matter:

This plant prefers soils balanced in organic matter and nutrients. Its presence can indicate fertile soil, often found in meadows, riverbanks, and pastures.

Preference for Slightly Acidic to Neutral Soils:

Rumex acetosa prefers soils with a pH range from slightly acidic to neutral, between 4.5 and 7.0. Its presence can indicate soil with a stable and often slightly acidic pH, suitable for many plants in meadows and wet environments.

Preference for Sunny to Partially Shaded Environments:

Common sorrel grows well in sunny to partially shaded environments. Its presence can indicate open or semi-open areas with good light exposure in a cool climate.

Indication of Natural or Semi-Natural Habitats:

This plant is often found in relatively stable habitats, such as natural meadows, pastures, and riverbanks. Its presence can indicate a relatively stable and undisturbed ecosystem.

RUMEX OBTUSIFOLIUS (BROAD-LEAVED DOCK)

MOIST SOILS
SOILS SATURATED WITH ORGANIC MATTER
NITROGEN-RICH SOILS
OVERGRAZING, EXCESSIVE FERTILIZATION, AND SOIL DISTURBANCE
POLLUTED SOILS
SOIL PH OF 5.0 TO 7.0
OPEN OR SEMI-OPEN AREAS
DISTURBED HABITATS

Preference for Moist to Moderately Moist Soils:

Broad-leaved dock thrives in moist to moderately moist soils. Its presence often indicates soil that retains moisture and can be partially waterlogged.

Tolerance for Soils Rich in Organic Matter and Nutrients:

This plant prefers compacted soils saturated with organic matter and nutrients. It is often a consequence of overgrazing, excessive nitrogen fertilization, and soil disturbance in wet conditions. The soil is anaerobic, polluted, and rich in nitrates.

Preference for Neutral to Slightly Acidic Soils:

Rumex obtusifolius prefers soils with a pH range from slightly acidic to neutral, between 5.0 and 7.0. Its presence can indicate soil with a stable and generally acidic pH.

Preference for Sunny to Partially Shaded Environments:

Broad-leaved dock grows well in sunny to partially shaded environments. Its presence can indicate open or semi-open areas with good light exposure in a cool climate.

Tolerance to Disturbance:

Rumex obtusifolius is relatively tolerant to disturbance and can quickly colonize disturbed areas. Its presence can indicate habitats that have experienced natural or human disturbances, such as agricultural lands, gardens, and intensively grazed pastures.

SALIX ALBA (WHITE WILLOW)

MOIST TO WATERLOGGED SOILS
GROUNDWATER RESURGENCE
RICH SOILS
SOIL PH OF 5.5 TO 7.0
SUNNY ENVIRONMENTS
DISTURBED HABITATS

Preference for Moist to Waterlogged Soils:

White willow thrives in moist to waterlogged soils. Its presence often indicates soil that retains a lot of moisture, commonly found in flood-prone areas, riverbanks, and marshes. It can also signal the presence of a groundwater resurgence.

Tolerance for Soils Rich in Organic Matter:

This plant prefers soils rich in bases, plant, and animal organic matter. Its presence can indicate fertile soil, often found in wet environments with an accumulation of nutrients.

Preference for Slightly Acidic to Slightly Basic Soils:

Salix alba prefers soils with a pH range from slightly acidic to slightly basic, between 5.5 and 7.5.

Preference for Sunny Environments:

White willow grows well in sunny environments in cool climates. Its presence can indicate open areas with little competition from other tree species, as it is a pioneer species.

Tolerance to Disturbance:

Salix alba is relatively tolerant to disturbance and can quickly colonize disturbed areas. Its presence can indicate habitats that have experienced natural or human disturbances.

SALIX CAPREA (GOAT WILLOW)

MOIST SOILS
NITROGEN-RICH SOILS
SOIL PH OF 5.0 TO 7.0+
SUNNY ENVIRONMENTS
DISTURBED HABITATS

Preference for Well-Drained to Moist Soils:

Goat willow thrives in well-drained to moist soils. Its presence often indicates soil that retains sufficient moisture and can sometimes be waterlogged.

Tolerance for Soils Rich in Organic Matter and Nutrients:

This plant prefers soils rich in organic matter and nutrients, particularly nitrogen. Its presence can indicate fertile soil, often found in wet meadows, riverbanks, ditches, and wooded areas.

Preference for Neutral to Slightly Acidic Soils:

Salix caprea prefers soils with a pH range from slightly acidic to neutral, between 5.0 and 7.0. However, it can tolerate more acidic or more alkaline conditions.

Preference for Sunny to Partially Shaded Environments:

Goat willow grows well in sunny environments without competition from other tree species (pioneer tree). Its presence can indicate open areas in a cool climate.

Tolerance to Disturbance:

Salix caprea is highly tolerant to disturbance, and its pioneer nature allows it to quickly colonize disturbed areas. Its presence can indicate habitats that have experienced natural or human` disturbances, such as roadsides, clearings, and deforested areas.

SALVIA PRATENSIS (MEADOW SAGE)

MODERATELY DRY SOILS
POOR SOILS
SOIL EROSION
NITROGEN-DEFICIENT SOILS
SOIL PH OF 6.0 TO 8.0
OPEN AREAS
PERIODS OF LOW WATER AVAILABILITY

Preference for Well-Drained to Moderately Dry Soils:
Meadow sage thrives in well-drained to moderately dry soils. Its presence often indicates soil that does not retain excessive water, allowing good root aeration.

Tolerance for Poor to Moderately Nutrient-Rich Soils:
This plant can develop in soils that are poor to moderately rich in nutrients and particularly deficient in animal organic matter. Its presence can indicate soil undergoing erosion that is not particularly fertile, often found in dry meadows, grassy slopes, and calcareous lawns.

Preference for Neutral to Slightly Alkaline Soils:
Salvia pratensis prefers soils with a pH range from neutral to slightly alkaline, between 6.0 and 8.0. Its presence can indicate soil with a stable and often slightly basic pH.

Preference for Sunny Environments:
Meadow sage grows well in environments with full sun exposure. Its presence can indicate open areas with full light exposure in a warm climate.

Tolerance to Drought:
Salvia pratensis is relatively drought-resistant. Its presence can indicate environments that may experience periods of drought or have low water availability.

SAMBUCUS EBULUS (DWARF ELDER)

MOIST SOILS
BASE-RICH SOILS
NITROGEN-RICH SOILS
POLLUTED SOILS
DEEP SOILS
PH OF 5.0 TO 7.0
OPEN OR SEMI-OPEN AREAS
DISTURBED HABITATS

Preference for Soils Rich in Organic Matter and Nutrients:
Dwarf elder thrives in soils rich in bases, organic matter, and nutrients. Its presence may indicate deep, moist soil rich in organic matter, particularly of animal origin, often found in disturbed areas, fallow lands, roadsides, and wastelands. If no visible organic matter is present but there is a nearby watercourse, it may indicate it is polluted by nitrates. This plant likes to grow near cultivated fields polluted by fertilizers and agricultural practices.

Preference for Well-drained to Moderately Moist Soils:
This plant prefers moderately moist and cool soils. Its presence often indicates deep soil that retains moisture well but is not constantly waterlogged. Water availability throughout the year is very satisfactory.

Preference for Neutral to Slightly Acidic Soils:
Sambucus ebulus prefers soils with a pH ranging from neutral to slightly acidic, i.e., a pH range of 5.0 to 7.0.

Preference for Sunny to Partially Shaded Environments:
Dwarf elder thrives in sunny to partially shaded environments.
Its presence can indicate open or semi-open areas with good light exposure in cool climates.

Tolerance to Disturbances:
Sambucus ebulus is relatively tolerant of disturbances and can quickly colonize disturbed areas. Its presence may indicate a habitat that has undergone natural or human disturbances, such as roadsides, fallow lands, and abandoned fields.

SAMBUCUS NIGRA (ELDERBERRY)

MOIST SOILS
RICH SOILS
SPRING RESURGENCE
PROXIMITY TO RECENT OR ANCIENT HUMAN HABITATION
SOIL PH OF 6.0 TO 8.0
OPEN OR SEMI-OPEN AREAS
DISTURBED HABITATS

Preference for Well-Drained to Moist Soils:

Elderberry thrives in moist soils. Its presence often indicates soil that retains sufficient moisture without being constantly waterlogged. It can indicate an underground spring or spring resurgence.

Tolerance for Soils Rich in Organic Matter:

This plant, often found near human habitation, prefers deep soils rich in bases, organic matter, and nutrients. Its presence can indicate fertile, nitrogen-rich soil, commonly found near human settlements, in hedgerows, forest edges, fallow lands, and along watercourses.

Preference for Neutral to Slightly Alkaline Soils:

Sambucus nigra prefers soils with a pH range from neutral to slightly alkaline, between 6.0 and 8.0. Its presence can indicate soil with a stable and often slightly basic pH.

Preference for Sunny to Partially Shaded Environments:

Elderberry grows well in sunny to partially shaded environments. Its presence can indicate open or semi-open areas with good light exposure in a cool climate.

Tolerance to Disturbance:

Sambucus nigra is relatively tolerant to disturbance and can quickly colonize disturbed areas, particularly those affected by human activity.

SANGUISORBA MINOR (SALAD BURNET)

DRY SOILS
SOILS RICH IN BASES
SOILS DEFICIENT IN ANIMAL ORGANIC MATTER
SOIL PH OF 6.0 TO 8.0
SUNNY ENVIRONMENTS
HYDRIC CONTRASTS AND DROUGHT

Preference for Well-Drained to Dry Soils:
Salad burnet thrives in well-drained to dry soils. Its presence often indicates soil subject to strong hydric contrasts and poor water retention, allowing good root aeration.

Tolerance for Poor to Moderately Nutrient-Rich Soils:
This plant can develop in soils rich in bases but poor in nutrients. Its presence can indicate soil deficient in animal organic matter, often found in dry lawns, calcareous meadows, and roadsides.

Preference for Neutral to Alkaline Soils:
Sanguisorba minor prefers soils with a pH range from neutral to alkaline, between 6.0 and 8.0. Its presence can indicate soil with a stable and often basic pH.

Preference for Sunny Environments:
Salad burnet grows well in environments with full sun exposure. Its presence can indicate open areas with full light exposure.

Tolerance to Drought:
Sanguisorba minor is relatively drought-resistant. Its presence can indicate environments that may experience periods of drought or have limited water availability.

SAPONARIA OFFICINALIS (SOAPWORT)

WELL-DRAINED TO MODERATELY MOIST SOILS
RICH SOILS
SOIL PH OF 6.0 TO 8.0
OPEN AREAS
DISTURBED HABITATS

Preference for Well-Drained to Moderately Moist Soils:

Soapwort thrives in well-drained to moderately moist soils. Its presence often indicates soil that retains sufficient moisture without being constantly waterlogged.

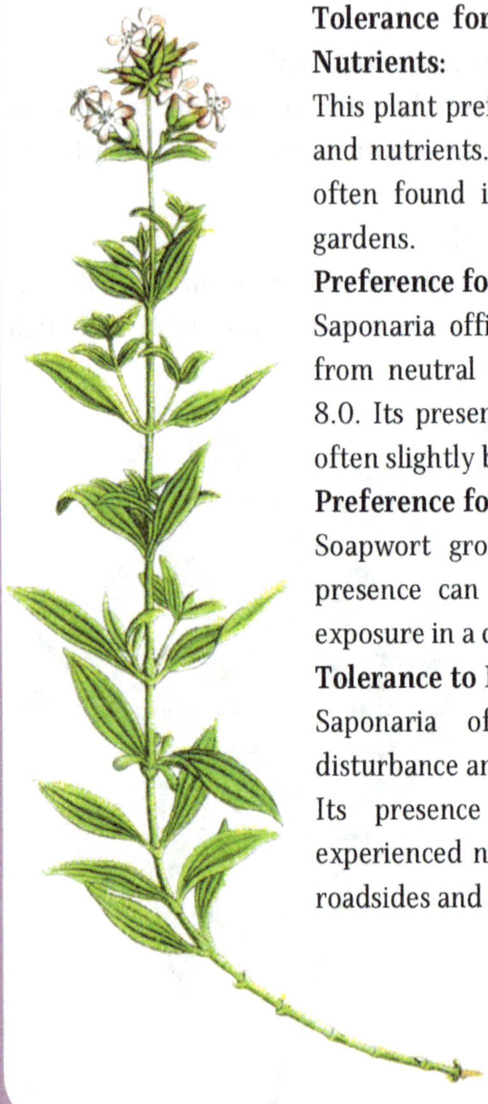

Tolerance for Soils Rich in Organic Matter and Nutrients:

This plant prefers soils rich in animal organic matter and nutrients. Its presence can indicate fertile soil, often found in meadows, riverbanks, ditches, and gardens.

Preference for Neutral to Slightly Alkaline Soils:

Saponaria officinalis prefers soils with a pH range from neutral to slightly alkaline, between 6.0 and 8.0. Its presence can indicate soil with a stable and often slightly basic pH.

Preference for Sunny Environments:

Soapwort grows well in sunny environments. Its presence can indicate open areas with good light exposure in a cool climate.

Tolerance to Disturbance:

Saponaria officinalis is relatively tolerant to disturbance and can quickly colonize disturbed areas. Its presence can indicate habitats that have experienced natural or human disturbances, such as roadsides and gardens.

SATUREJA MONTANA (WINTER SAVORY)

DRY SOILS
POOR SOILS
SOIL PH OF 6.5 TO 8.5
OPEN AREAS
DROUGHT CONDITIONS

Preference for Well-Drained to Dry Soils:

Winter savory thrives in well-drained to dry soils. Its presence often indicates soil that does not retain water, allowing good root aeration and avoiding water saturation.

Tolerance for Poor to Moderately Nutrient-Rich Soils:

This plant can develop in very thin soils, poor in organic matter and nutrients. Its presence can indicate soil that is not particularly fertile, often found in rocky terrains, cliffs, garrigues, and maquis.

Preference for Neutral to Alkaline Soils:

Satureja montana prefers soils with a pH range from neutral to alkaline, between 6.5 and 8.5. Its presence can indicate soil with a stable and often slightly basic pH.

Preference for Sunny Environments:

Winter savory grows well in sunny environments. Its presence can indicate open areas with full light exposure in a warm climate.

Tolerance to Drought:

Satureja montana is relatively drought-resistant. Its presence can indicate environments that may experience periods of drought or have low water availability. It can live in soils that are moist in winter but dry in summer.

SAXIFRAGA GRANULATA (MEADOW SAXIFRAGE)

WELL-DRAINED TO DRY SOILS
POOR SOILS
SOILS DEFICIENT IN ANIMAL ORGANIC MATTER
SOIL PH OF 5.0 TO 7.0
OPEN AREAS

Preference for Well-Drained to Moderately Dry Soils:

Meadow saxifrage thrives in well-drained to moderately dry soils. Its presence often indicates soil that does not retain excessive water, allowing good root aeration.

Tolerance for Poor to Moderately Nutrient-Rich Soils:

This plant can develop in soils poor to moderately rich in nutrients but often rich in bases. Its presence can indicate soil saturated with plant organic matter but deficient in animal organic matter, often found in poor meadows, dry lawns, clear-cuts, and roadsides.

Preference for Neutral to Slightly Acidic Soils:

Saxifraga granulata prefers soils with a pH range from slightly acidic to neutral, between 5.0 and 7.0. Its presence can indicate soil with a stable and often slightly acidic pH.

Preference for Sunny Environments:

Meadow saxifrage grows well in sunny environments in cool climates. Its presence can indicate open areas with good light exposure.

SCABIOSA COLUMBARIA (SMALL SCABIOUS)

DRY SOILS
POOR SOILS
ERODED SOILS
SOIL PH OF 6.0 TO 7.5
OPEN AREAS
PERIODS OF LOW WATER AVAILABILITY

Preference for Well-Drained to Dry Soils:
Small scabious thrives in very fine, well-drained to dry soils. Its presence often indicates soil that does not retain water.

Tolerance for Poor to Moderately Nutrient-Rich Soils:
This plant can develop in very eroded and poor to moderately nutrient-rich soils. Its presence can indicate rocky or sandy soil, often found in dry lawns, calcareous meadows, and fallow lands.

Preference for Neutral to Slightly Alkaline Soils:
Scabiosa columbaria prefers soils with a pH range from neutral to slightly alkaline, between 6.0 and 7.5. Its presence can indicate soil with a stable and often slightly basic pH.

Preference for Sunny Environments:
Small scabious grows well in environments with full sun exposure. Its presence can indicate open areas with full light exposure.

Tolerance to Drought:
Scabiosa columbaria is relatively drought-resistant. Its presence can indicate environments that may experience periods of drought or have limited water availability.

SEDUM ACRE (BITING STONECROP)

DRY SOILS
ROCKY OR SANDY SOILS
VERY POOR SOILS
SOIL PH OF 6.5 TO 8.0
SUNNY ENVIRONMENTS
DROUGHT CONDITIONS

Preference for Very Dry Soils:

Biting stonecrop thrives "sitting" (sedum in Latin means sitting) in cushion-like formations on stones, rocks, railway tracks, old roofs, or coarsely sandy soils. Its presence indicates rocky soil with almost no water retention capacity.

Tolerance for Nutrient-Poor Soils:

This plant can develop in nutrient-poor soils with almost no organic matter.

Preference for Neutral to Slightly Alkaline Soils:

Sedum acre prefers soils with a pH range from neutral to slightly alkaline, between 6.5 and 8.0. Its presence can indicate soil with a stable and often slightly basic pH.

Preference for Sunny Environments:

Biting stonecrop grows well in environments with full sun exposure. Its presence can indicate open areas with full light exposure.

Tolerance to Drought:

Sedum acre is highly drought-resistant. Its presence can indicate environments that may experience periods of drought or have limited water availability.

SENECIO JACOBAEA (COMMON RAGWORT)

WELL-DRAINED SOILS
SOILS RICH IN PLANT ORGANIC MATTER
SOILS DEFICIENT IN ANIMAL ORGANIC MATTER
SOIL PH OF 5.0 TO 7.0
OPEN OR SEMI-OPEN AREAS
DISTURBED HABITATS

Preference for Well-Drained to Moderately Moist Soils:

Common ragwort thrives in well-drained to moderately moist soils. Its presence often indicates soil that retains sufficient moisture without being constantly waterlogged.

Tolerance for Poor to Moderately Nutrient-Rich Soils:

This plant can develop in soils rich in plant organic matter and poor to deficient in animal organic matter. Its presence can indicate soil that is not particularly fertile and evolving towards forest conditions, often found in meadows, clearings, forest edges, roadsides, and waste grounds.

Preference for Slightly Acidic to Neutral Soils:

Senecio jacobaea prefers soils with a pH range from slightly acidic to neutral, between 5.0 and 7.0. Its presence can indicate soil with a stable and often slightly acidic pH.

Preference for Sunny to Partially Shaded Environments:

Common ragwort grows well in sunny to partially shaded environments. Its presence can indicate open or semi-open areas with good light exposure.

Tolerance to Disturbance:

Senecio jacobaea is relatively tolerant to disturbance and can quickly colonize disturbed areas. Its presence can indicate habitats that have experienced natural or human disturbances.

SILENE DIOICA (RED CAMPION)

MODERATELY MOIST SOILS
RICH SOILS
SOIL PH OF 5.5 TO 7.0
SEMI-SHADED ENVIRONMENTS
DISTURBED HABITATS

Preference for Well-Drained to Moist Soils:

Red campion thrives in well-drained to moist soils. Its presence often indicates soil with sand that retains sufficient moisture without being constantly waterlogged.

Tolerance for Soils Rich in Organic Matter and Nutrients:

This plant prefers soils rich in plant organic matter, nitrogen, and bases. Its presence can indicate fertile soil, often found in forest edges, meadows, hedgerows, riverbanks, and ditches.

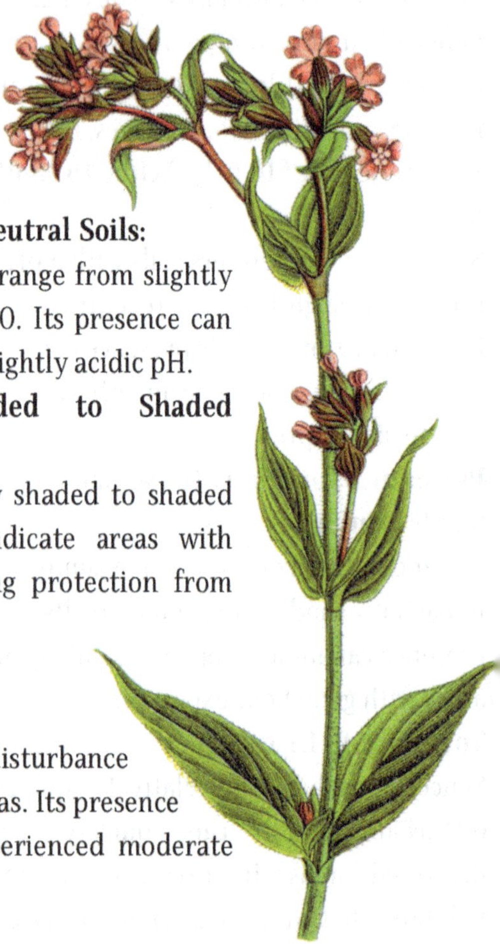

Preference for Slightly Acidic to Neutral Soils:

Silene dioica prefers soils with a pH range from slightly acidic to neutral, between 5.5 and 7.0. Its presence can indicate soil with a stable and often slightly acidic pH.

Preference for Partially Shaded to Shaded Environments:

Red campion grows well in partially shaded to shaded environments. Its presence can indicate areas with significant vegetation cover, offering protection from direct light in a cool climate.

Tolerance to Light to Moderate Disturbances:

Silene dioica is relatively tolerant to disturbance and can colonize lightly disturbed areas. Its presence can indicate habitats that have experienced moderate natural or human disturbances.

SILENE LATIFOLIA (WHITE CAMPION)

MODERATELY MOIST SOILS
RICH SOILS
SOIL PH OF 6.0 TO 8.0
SEMI-SHADED ENVIRONMENTS
DISTURBED HABITATS

Preference for Well-Drained to Moist Soils:

White campion thrives in well-drained to moist soils. Its presence often indicates soil that retains sufficient moisture without being constantly waterlogged, often sandy in texture.

Tolerance for Soils Rich in Organic Matter and Nutrients:

This plant prefers soils rich in plant organic matter, nitrogen, and bases. Its presence can indicate fertile soil, often found at forest edges, meadows, hedgerows, riverbanks, and ditches.

Preference for Neutral to Slightly Alkaline Soils:

Silene latifolia prefers soils with a pH range from neutral to slightly alkaline, between 6.0 and 8.0. Its presence can indicate soil with a stable and often slightly basic pH.

Preference for Partially Shaded to Shaded Environments:

White campion grows well in partially shaded to shaded environments. Its presence can indicate areas with significant vegetation cover, offering protection from direct light in a cool climate.

Tolerance to Light to Moderate Disturbances:

Silene latifolia is relatively tolerant to disturbance and can colonize lightly disturbed areas. Its presence can indicate habitats that have experienced moderate natural or human disturbances.

SINAPIS ARVENSIS (WILD MUSTARD)

MODERATELY MOIST SOILS
PHOSPHORUS BLOCKAGE
RECENT PH INCREASE
SOIL PH OF 6.5 TO 8.5
OPEN AREAS
DISTURBED HABITATS

Preference for Well-Drained to Moderately Moist Soils:

Wild mustard thrives in well-drained to moderately moist soils. Its presence often indicates soil that retains sufficient moisture without being constantly waterlogged.

Tolerance for Soils with Phosphorus Blockage:

This plant tolerates soils with phosphorus blockage. Its presence can indicate nutrient-rich soil with low phosphorus availability due to pH increase (e.g., from repeated lime application or stone grinding), often found in agricultural lands, cultivated fields, and gardens. It helps release phosphorus by acidifying the soil. It can, but should not be cultivated on acid soils.

Preference for Alkaline Soils:

Sinapis arvensis prefers soils with an alkaline pH range, between 6.5 and 8.5. Its presence can indicate a recent pH increase.

Preference for Sunny Environments:

Wild mustard grows well in sunny environments. Its presence can indicate open areas with full light exposure in a cool climate.

Tolerance to Disturbance:

Sinapis arvensis is relatively tolerant to disturbance and can quickly colonize disturbed areas. Its presence can indicate habitats that have experienced natural or human disturbances, often agricultural, such as cultivated fields, fallow lands, and roadsides.

SISYMBRIUM OFFICINALE (HEDGE MUSTARD)

WELL-DRAINED SOILS
SOILS RICH IN BASES
COMPACTED SOILS
PHOSPHORUS BLOCKAGE
RECENT PH INCREASE
SOIL PH OF 6.0 TO 8.0
OPEN AREAS
DISTURBED HABITATS

Preference for Well-Drained to Moderately Moist Soils:

Hedge mustard thrives in well-drained to moderately moist soils. Its presence often indicates soil that retains sufficient moisture without being constantly waterlogged.

Tolerance for Poor to Moderately Nutrient-Rich Soils:

This plant can develop in soils rich in bases with phosphorus blockage due to compaction and a rise in pH. Its presence can indicate soils affected by recent lime application or stone grinding.

Preference for Neutral to Slightly Alkaline Soils:

Sisymbrium officinale prefers soils with a pH range from neutral to slightly alkaline, between 6.0 and 8.0. Its presence can indicate a recent pH increase due to lime spreading or stone grinding.

Preference for Sunny Environments:

Hedge mustard grows well in sunny environments. Its presence can indicate open areas with full light exposure in a cool climate.

Tolerance to Disturbance:

Sisymbrium officinale is relatively tolerant to disturbance and can quickly colonize disturbed areas. Its presence can indicate habitats that have experienced natural or human disturbances, leading to soil compaction, such as fallow lands, agricultural areas, animal parks, gardens, roadsides, and waste grounds.

SOLANUM DULCAMARA (BITTERSWEET NIGHTSHADE)

MOIST SOILS
STRONG HYDRIC CONTRASTS
SOILS RICH IN BASES
SOILS RICH IN ANIMAL ORGANIC MATTER
SOIL PH OF 6.0 TO 8.0
SEMI-OPEN AREAS

Preference for Moist to Very Moist Soils:

Bittersweet nightshade thrives in humid to very humid soils. It is often found near watercourses, ditches, and marshy areas. It also tolerates drier soils, indicating strong hydric contrasts throughout the year.

Tolerance for Soils Rich in Organic Matter and Nutrients:

This plant prefers soils rich in bases and nitrogen. Its presence can indicate saturation with animal organic matter or an excess of nitrogen fertilizers, often found in wet areas, hedges, and riverbanks.

Preference for Neutral to Slightly Alkaline Soils:

Solanum dulcamara prefers soils with a pH range from neutral to slightly alkaline, between 6.0 and 8.0. Its presence can indicate soil with a stable and often slightly basic pH.

Preference for Sunny to Partially Shaded Environments:

Bittersweet nightshade grows well in sunny to partially shaded environments. Its presence can indicate open or semi-open areas with good light exposure in a cool climate.

SOLIDAGO VIRGAUREA (GOLDENROD)

DRY TO MOIST SOILS
POOR TO MODERATELY NUTRIENT-RICH SOILS
ROCKY SOILS
SOILS DEFICIENT IN ANIMAL ORGANIC MATTER
SOIL PH OF 5.0 TO 7.0
SEMI-SHADED ENVIRONMENTS
DISTURBED HABITATS

Preference for Dry to Moist Soils:

Goldenrod thrives in dry to moist soils. Its presence often indicates soil that retains sufficient moisture but experiences some hydric contrasts.

Tolerance for Poor to Moderately Nutrient-Rich Soils:

This plant can develop in soils that are poor to moderately rich in nutrients and bases. Its presence can indicate rocky soil saturated with plant organic matter but deficient in animal organic matter, often found in meadows, rubble, clearings, cuttings, forest edges, and roadsides.

Preference for Slightly Acidic to Neutral Soils:

Solidago virgaurea prefers soils with a pH range from slightly acidic to neutral, between 5.0 and 7.0. It can tolerate more acidic or basic soils.

Preference for Partially Shaded to Shaded Environments:

Goldenrod grows well in partially shaded to shaded environments. Its presence can indicate areas with significant vegetation cover, offering protection from direct light in a cool climate.

Tolerance to Disturbance:

Solidago virgaurea is relatively tolerant to disturbance and can quickly colonize disturbed areas. Its presence indicates habitats that have experienced natural or human disturbances.

SONCHUS ARVENSIS (PERENNIAL SOW-THISTLE)

MOIST SOILS
HYDRIC CONTRASTS
SOILS RICH IN BASES
COMPACTED SOILS
SALINIZATION
SOIL PH OF 6.0 TO 8.0
OPEN OR SEMI-OPEN AREAS
DISTURBED HABITATS

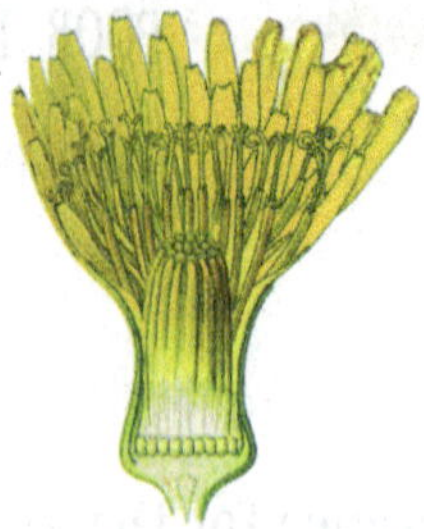

Preference for Well-Drained to Moist Soils:

Perennial sow-thistle thrives in well-drained to moist soils. Its presence often indicates soil that retains sufficient moisture and can be waterlogged, with hydric contrasts throughout the year.

Tolerance for Soils Rich in Organic Matter and Nutrients:

This plant indicates soil congested in bases and organic matter due to compaction and moisture. Its presence indicates soil that has been worked in wet conditions or has been naturally compacted, such as near river overflow. It shows some salt tolerance and may indicate irrigation in hot weather, causing causing the deposit of mineral salts from the evaporating water on the surface of the soil. It is often found in agricultural lands, riverbanks, meadows, and roadsides.

Preference for Neutral to Slightly Alkaline Soils:

Sonchus arvensis prefers soils with a pH range from neutral to slightly alkaline, between 6.0 and 8.0.
Its presence can indicate soil with a stable and often slightly basic pH.

Preference for Sunny to Partially Shaded Environments:

Perennial sow-thistle grows well in sunny to partially shaded environments. Its presence can indicate open or semi-open areas with good light exposure in a cool climate.

Tolerance to Disturbance:

Sonchus arvensis is relatively tolerant to disturbance and can quickly colonize disturbed areas.

SORBUS AUCUPARIA (ROWAN TREE)

WELL-DRAINED SOILS
HIGH ATMOSPHERIC HUMIDITY
POOR TO MODERATELY NUTRIENT-RICH SOILS
SOIL PH OF 4.5 TO 7.0, BUT CAN TOLERATE ABOVE 7.0
OPEN OR SEMI-OPEN AREAS

Preference for Well-Drained to Moist Soils:

Rowan tree thrives in well-drained to moist soils. Its presence often indicates soil that retains sufficient moisture without being constantly waterlogged. It requires high atmospheric humidity.

Tolerance for Poor to Moderately Nutrient-Rich Soils:

This plant can develop in a variety of soils, from poor to moderately rich in nutrients. Its presence generally indicates soil that is poor in bases and acidic. In mountainous areas, it is tolerant of a wide range of soils.

Preference for Acidic to Neutral Soils:

Sorbus aucuparia prefers soils with a pH range from slightly acidic to neutral, between 4.5 and 7.0. Its presence can indicate soil with a stable and often acidic pH. However, it also shows good tolerance to calcareous soils and is not incompatible with the presence of limestone.

Preference for Sunny to Partially Shaded Environments:

Rowan tree grows well in sunny to partially shaded environments. Its presence can indicate open or semi-open areas with good light exposure in a cool climate.

SPIRAEA ALBA (WHITE MEADOWSWEET)

MOIST SOILS
RICH SOILS
SATURATION WITH ORGANIC MATTER
SOIL PH OF 5.0 TO 8.5
OPEN OR SEMI-OPEN AREAS
DISTURBED HABITATS

Preference for Well-Drained to Moderately Moist Soils:
White meadowsweet thrives in well-drained to moderately moist soils. Its presence often indicates soil that retains sufficient moisture without being constantly waterlogged.

Tolerance for Soils Rich in Organic Matter and Nutrients:
This plant prefers rocky soils rich in bases and very rich in organic matter and nutrients. Its presence can indicate soil saturated with organic matter, often found in wet meadows, riverbanks, ditches, and marshy areas.

Preference for Slightly Acidic to Alkaline Soils:
Spiraea alba prefers soils with a pH range from slightly acidic to alkaline, between 5.0 and 8.5. Its presence can indicate soil with a stable and often basic pH.

Preference for Sunny to Partially Shaded Environments:
White meadowsweet grows well in sunny to partially shaded environments. Its presence can indicate open or semi-open areas with good light exposure in a cool climate.

Tolerance to Light to Moderate Disturbances:
Spiraea alba is relatively tolerant to disturbance and can colonize lightly disturbed areas. Its presence can indicate habitats that have experienced moderate natural or human disturbances.

STACHYS OFFICINALIS (BETONY)

WELL-DRAINED SOILS
MODERATELY RICH SOILS
SOIL PH OF 5.5 TO 7.0
OPEN OR SEMI-OPEN AREAS

Preference for Well-Drained to Moderately Moist Soils:
Betony thrives in well-drained to moderately moist soils. Its presence often indicates soil that retains sufficient moisture without being constantly waterlogged.

Tolerance for Moderately Rich Soils:
This plant prefers soils moderately rich in bases, organic matter, and nutrients. Its presence can indicate moderately fertile soil, often found in meadows, clearings, and roadsides.

Preference for Neutral to Slightly Acidic Soils:
Stachys officinalis prefers soils with a pH range from slightly acidic to neutral, between 5.5 and 7.0. Its presence can indicate soil with a stable and often slightly acidic pH.

Preference for Sunny to Partially Shaded Environments:
Betony grows well in sunny to partially shaded environments. Its presence can indicate open or semi-open areas with good light exposure in a cool climate.

STELLARIA HOLOSTEA (GREATER STITCHWORT)

MODERATELY MOIST SOILS
RICH SOILS
HEALTHY AND BALANCED SOILS
SOIL PH OF 5.0 TO 7.0
PARTIALLY SHADED ENVIRONMENTS

Preference for Well-Drained to Moderately Moist Soils:

Greater stitchwort thrives in well-drained to moderately moist soils. Its presence often indicates soil that retains sufficient moisture without being constantly waterlogged.

Tolerance for Soils Rich in Organic Matter and Nutrients:

This plant prefers soils rich in bases, organic matter, and nutrients. Its presence can indicate fertile and healthy soil, balanced in nutrients and water, often found in woodlands, hedgerows, and forest edges.

Preference for Neutral to Slightly Acidic Soils:

Stellaria holostea prefers soils with a pH range from slightly acidic to neutral, between 5.0 and 7.0. Its presence can indicate soil with a stable and often slightly acidic pH, suitable for many plants. It also tolerates more acidic or basic pH levels.

Preference for Shaded to Partially Shaded Environments:

Greater stitchwort grows well in shaded to partially shaded environments. Its presence can indicate semi-shaded to shaded areas with good vegetative cover in a cool climate.

STELLARIA MEDIA (COMMON CHICKWEED)

MODERATELY MOIST SOILS
RICH SOILS
HEALTHY AND BALANCED SOILS
SOIL PH OF 5.5 TO 7.0
SEMI-SHADED TO OPEN AREAS

Preference for Well-Drained to Moderately Moist Soils:

Common chickweed thrives in well-drained to moderately moist soils. Its presence often indicates soil that retains sufficient moisture without being constantly waterlogged.

Tolerance for Soils Rich in Organic Matter and Nutrients:

This plant prefers healthy soils rich in organic matter and nutrients. Its presence can indicate fertile and balanced soil, often found in gardens, cultivated fields, meadows, and roadsides.

Preference for Neutral to Slightly Acidic Soils:

Stellaria media prefers soils with a pH range from slightly acidic to neutral, between 5.5 and 7.0. Its presence can indicate soil with a stable pH, often slightly acidic or slightly basic.

Preference for Partially Shaded to Sunny Environments:

Common chickweed grows well in partially shaded to sunny environments. Its presence can indicate semi-shaded to open areas with good light exposure in a cool climate.

Tolerance to Disturbance:

Stellaria media is highly tolerant to disturbance and can quickly colonize disturbed areas. Its presence can indicate habitats that have experienced natural or human disturbances, such as gardens, cultivated fields, and roadsides.

SYMPHYTUM OFFICINALE (COMMON COMFREY)

HUMID TO WATERLOGGED SOILS
RICH SOILS
SOIL PH OF 5.5 TO 7.0
OPEN OR SEMI-OPEN AREAS
DISTURBED HABITATS

Preference for Humid to Waterlogged Soils:

Common comfrey thrives in humid to waterlogged soils. Its presence often indicates soil saturated with water, commonly found near watercourses, ditches, and marshy areas. It can tolerate hydric contrasts throughout the year.

Tolerance for Soils Rich in Organic Matter and Nutrients:

This plant prefers soils rich in bases, organic matter, and nutrients, particularly nitrogen. Its presence can indicate fertile soil, often found in wet meadows, riverbanks, and ditches.

Preference for Neutral to Slightly Acidic Soils:

Symphytum officinale prefers soils with a pH range from slightly acidic to neutral, between 5.5 and 7.0. Its presence can indicate soil with a stable and often slightly acidic pH.

Preference for Sunny to Partially Shaded Environments:

Common comfrey grows well in sunny to partially shaded environments. Its presence can indicate open or semi-open areas with good light exposure in a cool climate.

Tolerance to Disturbance:

Symphytum officinale is relatively tolerant to disturbance and can quickly colonize disturbed areas. Its presence can indicate habitats that have experienced natural or human disturbances, such as agricultural lands, wet meadows, and riverbanks.

TANACETUM VULGARE (TANSY)

WELL-DRAINED TO MODERATELY MOIST SOILS
POOR TO RICH SOILS
METAL-POLLUTED SOILS
SOIL PH OF 4 TO 7.5
OPEN OR SEMI-OPEN AREAS
HIGHLY DISTURBED HABITATS

Preference for Well-Drained to Moderately Moist Soils:

Tansy thrives in well-drained to moderately moist soils. Its presence often indicates soil that retains sufficient moisture without being constantly waterlogged.

Tolerance for Metal-Polluted Soils:

This plant can develop in soils poor to moderately rich in nutrients. Its presence can indicate surface soil polluted by metals, especially if it is very abundant. Germination improves in metal-contaminated surface soils, often found on former lignite deposits and poor acidic soils of old mines. It is resistant to many metallic elements (Hg, Pb, Cd, Ni, Cr, Zn, and Mn). Its roots concentrate cadmium, chromium, lead, and iron, while its leaves contain cadmium, zinc, and magnesium.

Preference for Neutral to Slightly Acidic Soils:

Tanacetum vulgare prefers soils with a pH range from acidic to neutral, between 4 and 7.5.

Preference for Sunny to Partially Shaded Environments:

Tansy grows well in sunny to partially shaded environments. Its presence can indicate open or semi-open areas with good light exposure in a cool climate.

Tolerance to Disturbance:

Tanacetum vulgare is highly tolerant to disturbance and can quickly colonize disturbed areas. Its presence generally indicates habitats that have experienced natural or human disturbances.

TARAXACUM OFFICINALE (DANDELION)

WELL-DRAINED SOILS
SOILS RICH IN ANIMAL ORGANIC MATTER
SOIL PH OF 5.0 TO 7.0
OPEN AREAS
DISTURBED HABITATS

Preference for Well-Drained to Moderately Moist Soils:

Dandelion thrives in well-drained to moderately moist soils. Its presence often indicates soil that retains sufficient moisture without being constantly waterlogged.

Tolerance for Soils Rich in Organic Matter:

This plant prefers soils rich in bases and animal organic matter. Its presence can indicate nutrient-rich soil that may be becoming nitrogen-saturated and slightly compacted, often due to agricultural practices or grazing.

Preference for Neutral to Slightly Acidic Soils:

Taraxacum officinale prefers soils with a pH range from neutral to slightly alkaline, between 5.0 and 7.0. Its presence can indicate soil with a stable pH suitable for many plants in disturbed environments.

Preference for Sunny Environments:

Dandelion grows well in sunny environments. Its presence can indicate open areas with good light exposure in a cool climate.

Tolerance to Disturbance:

Taraxacum officinale is highly tolerant to disturbance and can quickly colonize disturbed areas. Its presence indicates habitats that have experienced natural or human disturbances, such as lawns, meadows, gardens, and wastelands.

TAXUS BACCATA (ENGLISH YEW)

WELL-DRAINED SOILS
SOILS VERY RICH IN BASES
POSSIBLE PRESENCE OF LIMESTONE
SOIL PH OF 6.0 TO 8.0
SEMI-SHADED OR SHADED AREAS
DENSE FORESTS
STABLE ECOSYSTEM

Preference for Well-Drained to Moderately Moist Soils:

English yew thrives in well-drained to moderately moist soils. Its presence often indicates soil that retains sufficient moisture without being constantly waterlogged. It also requires very high atmospheric humidity and often grows in association with beech, which demands similar conditions.

Tolerance for Soils Rich in Organic Matter:

This plant prefers soils very rich in bases. It tolerates rocky and calcareous soils well, indicating the presence of limestone in the soil.

Preference for Neutral to Slightly Alkaline Soils:

Taxus baccata prefers soils with a pH range from neutral to slightly alkaline, between 6.0 and 8.0. Its presence can indicate soil with a stable and often slightly basic pH.

High Shade Tolerance:

English yew grows well in partially shaded to shaded environments. It can live its entire life in the shade, making it indicative of dense forests and shaded habitats. Its presence can indicate semi-shaded or shaded areas with dense vegetative cover in a cool climate.

Indication of Natural or Semi-Natural Habitats:

This slow-growing tree can be found in relatively stable habitats, such as ancient beech forests, pubescent oak forests, linden forests, or other dense and humid forests, often in less accessible areas. Its presence can indicate a relatively stable and undisturbed ecosystem.

THYMUS SERPYLLUM (WILD THYME)

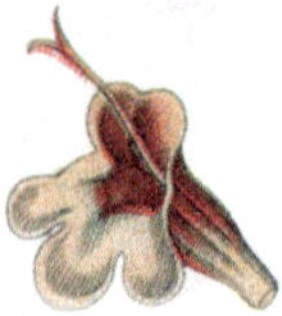

DRY SOILS
SOILS RICH IN BASES
POOR NUTRIENT SOILS
ABSENCE OF HUMUS
PERIODS OF DROUGHT
SOIL PH OF 6.0 TO 8.0
OPEN AREAS

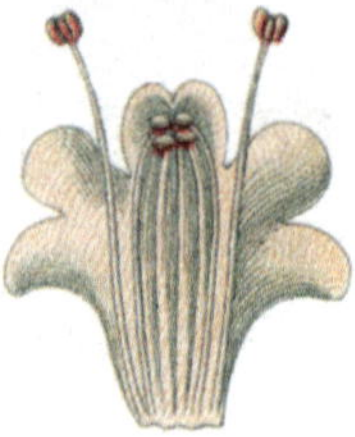

Preference for Well-Drained to Dry Soils:
Wild thyme thrives in well-drained to dry soils. Its presence often indicates rocky or sandy soil that does not retain water and lacks humus. Thymus serpyllum is relatively drought-resistant, indicating environments that can experience periods of drought or have low water availability.

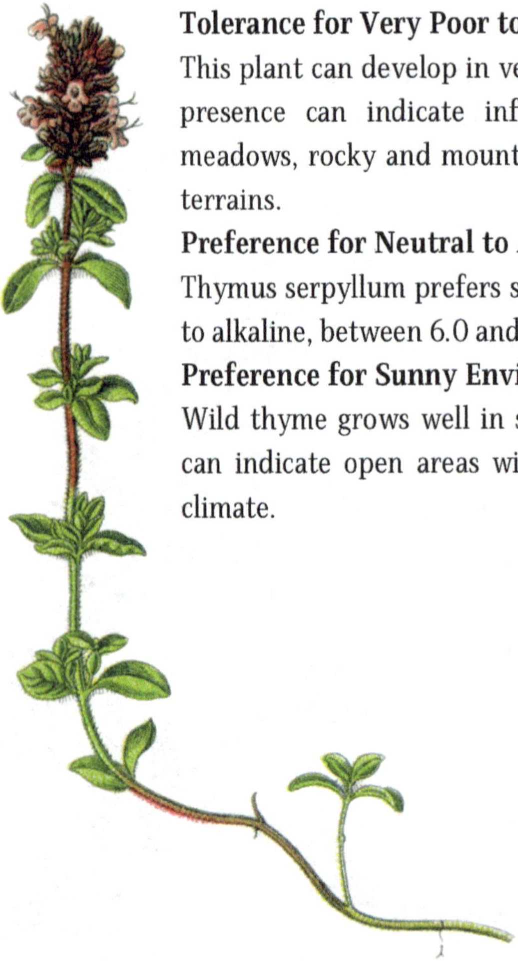

Tolerance for Very Poor to Poor Nutrient Soils:
This plant can develop in very poor to poor nutrient soils. Its presence can indicate infertile soil, often found in dry meadows, rocky and mountainous grasslands, and calcareous terrains.

Preference for Neutral to Alkaline Soils:
Thymus serpyllum prefers soils with a pH range from neutral to alkaline, between 6.0 and 8.0.

Preference for Sunny Environments:
Wild thyme grows well in sunny environments. Its presence can indicate open areas with full light exposure in a warm climate.

THYMUS VULGARIS (COMMON THYME)

DRY SOILS
SOILS RICH IN BASES
POOR NUTRIENT SOILS
ABSENCE OF HUMUS
PERIODS OF DROUGHT
SOIL PH OF 6.0 TO 8.0
OPEN AREAS

Preference for Well-Drained to Dry Soils:

Common thyme thrives in well-drained to dry soils. Its presence often indicates rocky or sandy soil that does not retain water and lacks humus. Thymus vulgaris is relatively drought-resistant, indicating environments that can experience periods of drought or have low water availability.

Tolerance for Very Poor to Poor Nutrient Soils:

This plant can develop in very poor to poor nutrient soils. Its presence can indicate infertile soil, often found in garrigues, maquis, and rocky terrains.

Preference for Neutral to Alkaline Soils:

Thymus vulgaris prefers soils with a pH range from neutral to alkaline, between 6.0 and 8.0. Its presence can indicate soil with a stable, often slightly basic, pH.

Preference for Sunny Environments:

Common thyme grows well in sunny environments. Its presence can indicate open areas with full light exposure in a warm climate.

TILIA CORDATA (SMALL-LEAVED LIME)

MOIST SOILS
HIGH ATMOSPHERIC HUMIDITY
RICH SOILS
SOIL PH OF 5.0 TO 7.0
SEMI-OPEN AREAS
DISTURBED HABITATS

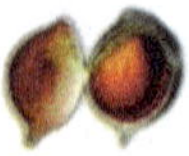

Preference for Well-Drained to Moderately Moist Soils:

The small-leaved lime tree thrives in well-drained to moderately moist soils. Its presence often indicates soil that retains sufficient moisture without being constantly waterlogged. It requires and indicates high atmospheric humidity, and in warm regions, its presence indicates moist soil.

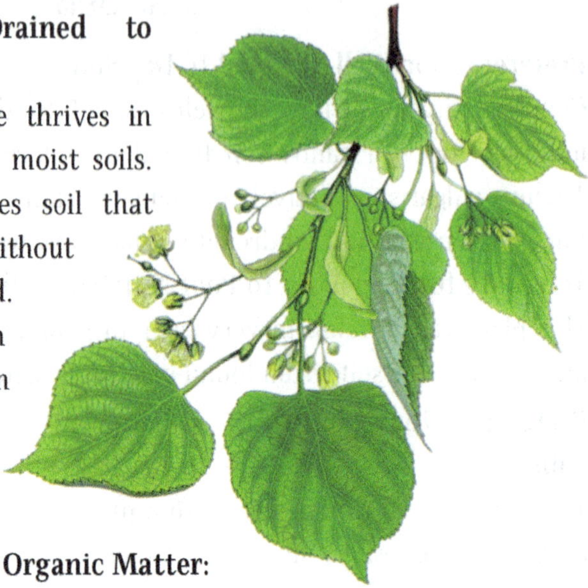

Tolerance for Soils Rich in Organic Matter:

This plant prefers soils rich in bases, organic matter, and nutrients. Its presence can indicate fresh, deep soil, often found in deciduous forests, forest edges, and parks.

Preference for Neutral to Slightly Acidic Soils:

Tilia cordata prefers soils with a pH range from slightly acidic to neutral, between 5.0 and 7.0. Its presence can indicate soil with a stable pH suitable for many forest plants.

Preference for Sunny to Partially Shaded Environments:

The small-leaved lime tree grows well in sunny to partially shaded environments. Its presence can indicate open or semi-open areas with partial light exposure in a cool climate.

Tolerance to Disturbance:

Tilia cordata is relatively tolerant to disturbance and can colonize disturbed areas. Its presence can indicate habitats that have experienced natural or human disturbances, such as parks and roadsides.

TRAGOPOGON PRATENSIS (MEADOW SALSIFY)

MODERATELY MOIST SOILS
SOILS RICH IN BASES
HEALTHY AND BALANCED SOILS
SOIL PH OF 6.0 TO 8.0
OPEN OR SEMI-OPEN AREAS

Preference for Well-Drained to Moderately Moist Soils:
Meadow salsify thrives in well-drained to moderately moist soils. Its presence often indicates soil that retains sufficient moisture without being constantly waterlogged.

Tolerance for Moderately Nutrient-Rich Soils:
This plant prefers soils rich in bases and moderately rich in nutrients. Its presence can indicate healthy and balanced soil, often found in meadows, pastures, and roadsides.

Preference for Neutral to Slightly Alkaline Soils:
Tragopogon pratensis prefers soils with a pH range from neutral to slightly alkaline, between 6.0 and 8.0. Its presence can indicate soil with a stable pH, often slightly basic.

Preference for Sunny to Partially Shaded Environments:
Meadow salsify grows well in sunny to partially shaded environments. Its presence can indicate open or semi-open areas with good light exposure in a cool climate.

TRIFOLIUM PRATENSE (RED CLOVER)

MODERATELY MOIST SOILS
HEAVY SOILS
RICH SOILS
ABSENCE OF DROUGHT PERIODS
SOIL PH OF 6.0 TO 7.5
RECENT SOIL ACIDIFICATION
OPEN AREAS
DISTURBED HABITATS

Preference for Well-Drained to Moderately Moist Soils:

Red clover thrives in well-drained to moderately moist soils. Its presence often indicates soil that retains sufficient moisture without being constantly waterlogged. It does not tolerate drought well.

Tolerance for Soils Rich in Organic Matter:

This plant prefers soils rich in organic matter and nutrients. Its presence can indicate fertile soil, often found in meadows, pastures, and roadsides.

Preference for Neutral to Slightly Basic Soils:

Trifolium pratense prefers soils with a pH range from slightly basic to neutral, between 6.0 and 7.5. Its presence can indicate soil with a pH in the process of acidifying (pH drop) or having recently acidified.

Preference for Sunny Environments:

Red clover grows well in sunny environments. Its presence can indicate open areas with full light exposure in a cool climate.

Tolerance to Disturbance:

Trifolium pratense is relatively tolerant to disturbance and can quickly colonize disturbed areas. Its presence can indicate habitats that have experienced natural or human disturbances.

TUSSILAGO FARFARA (COLTSFOOT)

HEAVY AND MOIST SOILS
SOILS RICH IN BASES
SOILS MODERATELY RICH IN NUTRIENTS
PH RANGE OF 5.0 TO 7.0
OPEN AREAS
HIGHLY DISTURBED HABITATS

Preference for Moist to Well-Drained Soils:
Coltsfoot thrives in heavy, fresh, and moist soils. Its presence often indicates soil that retains sufficient moisture without being constantly waterlogged. It often suggests the presence of a nearby water pocket close to the surface.
Tolerance for Unstable Soils Moderately Rich in Nutrients:
This plant can develop in soils rich in bases and moderately rich in nutrients. Its presence indicates an unstable, disturbed, or even runoff-prone soil, preceding the colonization by pioneer trees in wet areas. It is found in moist forest edges, riverbanks, embankments, and disturbed soils.
Preference for Slightly Acidic to Neutral Soils:
Tussilago farfara prefers soils with a pH range of slightly acidic to neutral, between 5.0 and 7.0. Its presence can indicate soil with a stable pH, suitable for many plants in open and disturbed habitats.

Preference for Sunny Environments:
Coltsfoot grows well in environments with good sun exposure. Its presence can indicate open areas with full light exposure in a cool climate.
Tolerance to Disturbance:
Tussilago farfara is very tolerant to disturbance and can rapidly colonize disturbed areas. Its presence indicates a habitat that has experienced and continues to experience natural or human disturbances.

ULMUS GLABRA (WYCH ELM)

MOIST SOILS
SOILS RICH IN BASES
BALANCED SOILS
PH RANGE OF 5.0 TO 7.5
SEMI-SHADED AREAS

Preference for Well-Drained to Moist Soils:

Wych elm thrives in well-drained to moist soils. Its presence often indicates soil that retains enough moisture without being constantly waterlogged. It requires high atmospheric humidity.

Tolerance for Soils Rich in Organic Matter:

This plant prefers soils rich in bases and moderately rich in organic matter and nutrients. Its presence can indicate a balanced soil, often found in cool hill or mountain forests, river valleys, and wooded areas.

Preference for Slightly Acidic to Slightly Basic Soils:

Ulmus glabra prefers soils with a pH range from slightly acidic to slightly basic, between 5.0 and 7.5. Its presence can indicate soil with a stable pH, suitable for many forest plants. Ulmus glabra shows good tolerance for calcareous soils, and can thus indicate the presence of limestone in the soil.

Preference for Partially Shaded Environments:

Wych elm grows well in partially shaded environments. Its presence can indicate semi-shaded areas in a cool climate.

URTICA DIOICA (COMMON NETTLE)

MOIST SOILS
SOILS RICH IN BASES
SOILS RICH IN NITROGEN AND PHOSPHORUS
PRESENCE OF IRON IN THE SOIL
BIRD PERCHES
PH OF 5.0 TO 7.0
OPEN OR SEMI-OPEN AREAS
DISTURBED HABITATS

Preference for Well-drained to Moist Soils:

Common nettle prefers moist soils. Its presence often indicates hydromorphisms and a soil that retains much moisture.

Preference for Soils Rich in Organic Matter:

This plant thrives in soils rich in organic matter and nutrients, particularly nitrogen, phosphorus, and iron. Its presence may indicate fertile soil, a change in the state of iron in the soil, remnants of old scrap metal, or the recent presence of birds. It is often found near human habitation, in dumps, composts, ditches, fresh woods, and bird perches.

Tolerance to Slightly Acidic to Neutral Soils:

Urtica dioica prefers soils with a pH ranging from slightly acidic to neutral, i.e., a pH range of 5.0 to 7.0. Its presence can indicate soil with stable pH, suitable for many plants in disturbed environments.

Preference for Sunny to Partially Shaded Environments:

Urtica dioica thrives in sunny to partially shaded environments. Its presence can indicate open or semi-open areas with good light exposure in cool climates.

Tolerance to Disturbance:

Urtica dioica is highly tolerant to disturbance and can quickly colonize disturbed areas. Its presence may indicate a habitat that has undergone natural or human disturbances, such as wastelands, dumps, and roadsides.

VACCINIUM MYRTILLUS (BILBERRY)

WELL-DRAINED SOILS
HIGH ATMOSPHERIC HUMIDITY
NITROGEN-POOR SOILS
SOILS RICH IN PLANT ORGANIC MATTER
PH OF 4.0 TO 5.5
SHADED ENVIRONMENTS
STABLE HABITATS

Preference for Well-drained to Moderately Moist Soils:

This plant prefers well-drained to moderately moist soils. Its presence often indicates soil that retains enough moisture without being constantly waterlogged but is well-drained. Bilberry particularly likes forest areas with high atmospheric humidity and only thrives under these conditions.

Tolerance to Nutrient-poor Soils:

Vaccinium myrtillus can develop in nutrient-poor soils, especially those very low in nitrogen. Its presence may indicate soil rich in plant organic matter but deficient in animal organic matter. This plant, adapted to infertile soils, is often found in heathlands, coniferous underwoods, and acidic grasslands.

Preference for Acidic Soils:

Bilberry grows well in acidic soils. Its presence often indicates soil with a low pH, generally in the range of 4.0 to 5.5. Its presence may indicate podzolic soil (soil leaching in cold and humid regions) or acidic forest soil.

Preference for Partially Shaded to Shaded Environments:

Bilberry grows well in partially shaded to shaded environments in cold climates. Its presence can indicate underwoods or forest edges with good vegetation cover. At high altitudes, however, this plant may appreciate growing in full sun.

Indication of Natural or Semi-natural Environments:

This plant is often found in relatively stable habitats, such as coniferous forest underwoods, heathlands, and acidic grasslands. Its presence may indicate a relatively stable and undisturbed ecosystem.

VERBASCUM THAPSUS (COMMON MULLEIN)

WELL-DRAINED SOILS
PERIODS OF DROUGHT
SOILS RICH IN BASES
SOILS POOR IN ANIMAL ORGANIC MATTER
SOILS RICH IN PLANT ORGANIC MATTER
PH OF 6.0 TO 7.5
SUNNY ENVIRONMENTS
RECENTLY HEAVILY DISTURBED HABITAT

Preference for Well-drained Soils:

Common mullein thrives in well-drained soils. Its presence often indicates soil that does not retain water well. It is relatively drought-resistant. Its presence may indicate environments that can experience periods of drought or have limited water availability.

Tolerance to Nutrient-poor to Moderately Rich Soils:

This plant can develop in soils rich in bases, saturated with plant organic matter, and poor in animal organic matter. Its presence, often at forest edges, can indicate a barren or recently exposed soil, often due to logging, storms, or fire. It can also be found in windthrows, fallows, old quarries, screes, embankment areas, wastelands, roadsides, and disturbed areas.

Preference for Neutral to Slightly Alkaline Soils:

Verbascum thapsus prefers soils with a pH ranging from neutral to slightly alkaline, i.e., a pH range of 6.0 to 7.5. Its presence can indicate soil with stable and often slightly basic pH.

Preference for Sunny Environments:

Common mullein thrives in well-exposed sunny environments. Its presence can indicate open areas, often recent, with full light exposure, even sun-scorched.

Tolerance to Disturbance:

Verbascum thapsus is highly tolerant to disturbance and can quickly colonize disturbed areas. One could even say that this plant depends on disturbance since its seeds can only germinate and it grows well in bare soil. Its presence precisely indicates a habitat that has recently undergone significant natural or human disturbances.

VERONICA CHAMAEDRYS (GERMANDER SPEEDWELL)

WELL-DRAINED SOILS
MODERATELY NUTRIENT-RICH AND BALANCED SOILS
PH OF 5.5 TO 7.0
OPEN OR SEMI-OPEN AREAS
DISTURBED HABITATS

Preference for Well-drained to Moderately Moist Soils:

Germander speedwell thrives in well-drained to moderately moist soils. Its presence often indicates soil that retains enough moisture without being constantly waterlogged.

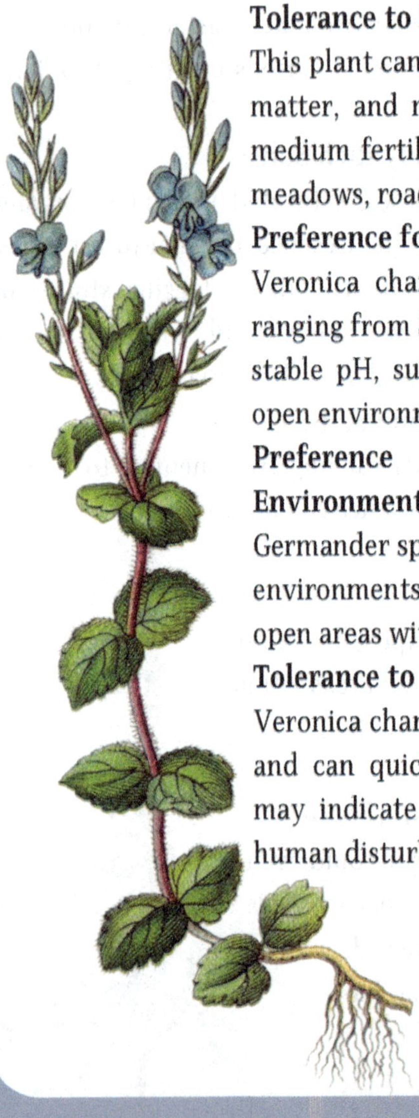

Tolerance to Moderately Nutrient-rich Soils:

This plant can develop in soils rich in bases, organic plant matter, and nitrogen. Its presence may indicate soil of medium fertility and relatively balanced, often found in meadows, roadsides, and forest edges.

Preference for Neutral Soils:

Veronica chamaedrys prefers soils with a neutral pH, ranging from 5.5 to 7.0. Its presence can indicate soil with stable pH, suitable for many plants in open and semi-open environments.

Preference for Sunny to Partially Shaded Environments:

Germander speedwell thrives in sunny to partially shaded environments. Its presence can indicate open or semi-open areas with good light exposure in cool climates.

Tolerance to Disturbance:

Veronica chamaedrys is relatively tolerant to disturbance and can quickly colonize disturbed areas. Its presence may indicate a habitat that has undergone natural or human disturbances, such as meadows and roadsides.

VICIA CRACCA (TUFTED VETCH)

DRY TO MODERATELY MOIST SOILS
SOILS RICH IN BASES
SOIL POOR TO RICH IN ORGANIC MATTER
COMPACTED SOILS
PH OF 5.5- TO 7.0+
OPEN OR SEMI-OPEN AREAS
DISTURBED HABITATS

Preference for Dry to Moderately Moist Soils:
Tufted vetch thrives in dry to moderately moist soils. Its presence often indicates soil that retains enough moisture without being overly wet.

Tolerance to Moderately Rich to Rich Soils:
This plant prefers soils rich in bases and ranging from poor to rich in organic matter and nutrients. Its presence may indicate compacted soil of medium to high fertility, often found in meadows, roadsides, hedgerows, and fallow lands. Like other legumes, it requires nitrogen to grow but lives in symbiosis with bacteria on its roots that absorb and make nitrogen from the air available. This explains why it thrives in both nitrogen-poor and nitrogen-rich soils and tolerates both well. It can become dominant in some nitrogen-poor meadows, having an advantage over surrounding plants, but its presence enriches the soil with nitrogen.

Preference for Neutral Soils:
Vicia cracca prefers soils with a neutral pH, ranging from 5.5 to 7.0. However, this plant is tolerant of more acidic or basic pH levels.

Preference for Sunny to Partially Shaded Environments:
Tufted vetch thrives in sunny to partially shaded environments. Its presence can indicate open or semi-open areas with good light exposure in cool climates.

Tolerance to Disturbances:
Vicia cracca is relatively tolerant to disturbance and can quickly colonize disturbed areas, particularly those compacted by soil work or heavy machinery (tractors). Its presence improves soil structure.

VICIA SATIVA (COMMON VETCH)

DRY TO MODERATELY MOIST SOILS
SOILS RICH IN BASES
SOIL FROM POOR TO RICH IN ORGANIC MATTER
COMPACTED SOILS
PH OF 5.5 TO 8.0
OPEN OR SEMI-OPEN AREAS
DISTURBED HABITATS

Preference for Well-drained to Moderately Moist Soils:
Common vetch thrives in well-drained to moderately moist soils. Its presence often indicates soil that retains enough moisture without being constantly waterlogged.

Tolerance to Moderately Rich to Rich Soils in Organic Matter and Nutrients:
This plant prefers soils rich in bases that range from poor to rich in organic matter and nutrients. Its presence may indicate compacted soil of medium to high fertility, often found in meadows, roadsides, hedgerows, and fallow lands. Like other legumes, it requires nitrogen to grow but lives in symbiosis with bacteria on its roots that absorb and make nitrogen from the air available. This explains why it thrives in both nitrogen-poor and nitrogen-rich soils and tolerates both well. It can become dominant in some nitrogen-poor meadows, having an advantage over surrounding plants, but its presence enriches the soil with nitrogen.

Preference for Neutral to Slightly Basic Soils:
Vicia sativa prefers soils with a pH ranging from neutral to slightly basic, i.e., a pH range of 5.5 to 8.0.

Preference for Sunny to Partially Shaded Environments:
Common vetch thrives in sunny to partially shaded environments. Its presence can indicate open or semi-open areas with good light exposure in cool climates.

Tolerance to Disturbances:
Vicia sativa is relatively tolerant to disturbance and can quickly colonize disturbed areas. Its presence may indicate a habitat that has undergone natural or human disturbances, such as cultivated fields, fallow lands, and roadsides. It often indicates soil compaction but improves soil structure by its presence.

VINCA MINOR (LESSER PERIWINKLE)

MOIST SOILS
BALANCED SOILS
LIGHT, NON-COMPACTED SOILS
INDICATOR OF ANCIENT HABITATS
PH OF 5.5 TO 8.5
SHADED TO PARTIALLY SHADED ENVIRONMENTS

Preference for Well-drained to Moderately Moist Soils:
Lesser periwinkle thrives in moderately moist soils. Its presence often indicates soil that retains moisture well without being waterlogged.

Tolerance to Soils Rich in Organic Matter:
This plant prefers soils rich in bases and moderately rich in organic matter and nutrients. Its presence indicates balanced soil, often found in underwoods, hedgerows, and gardens. It indicates light soil and does not tolerate compaction.

Archaeophyte Plant:
This plant has a medicinal and magical reputation dating back to ancient times. It grows on old graves and ancient archaeological sites and former habitation sites, particularly in Europe, as many studies explain, and its durability is exceptional. Due to its lack of tolerance for soil compaction, it does not withstand the modernization of ancient areas well and is often only apparent in abandoned sites and ruins, often in forests.

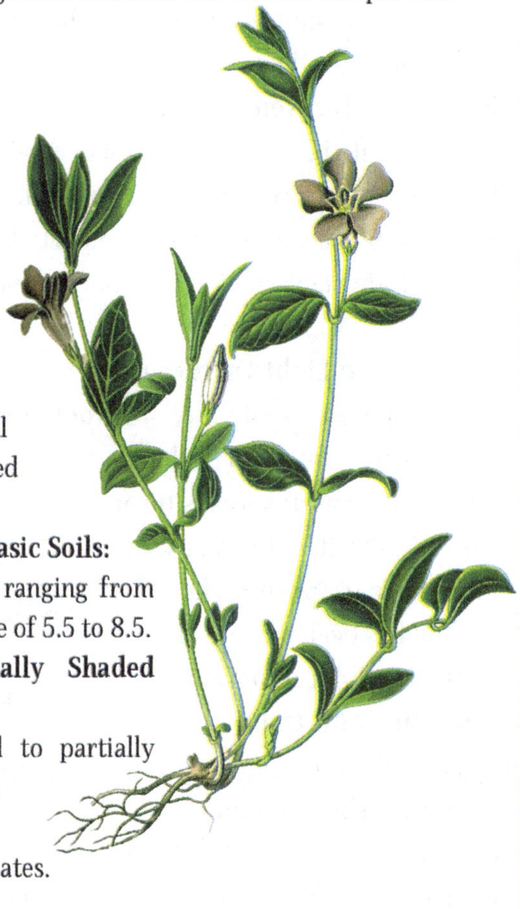

Preference for Neutral to Slightly Basic Soils:
Vinca minor prefers soils with a pH ranging from neutral to slightly basic, i.e., a pH range of 5.5 to 8.5.

Preference for Shaded to Partially Shaded Environments:
Lesser periwinkle thrives in shaded to partially shaded environments. Its presence can indicate semi-shaded to shaded areas with good vegetative cover in cool climates.

VIOLA ODORATA (SWEET VIOLET)

MOIST SOILS
RICH SOILS
PH OF 5.5 TO 7.0
PARTIALLY SHADED ENVIRONMENTS
LIGHTLY DISTURBED HABITATS

Preference for Well-drained to Moderately Moist Soils:
Sweet violet thrives in well-drained to moderately moist soils. Its presence often indicates soil that retains enough moisture without being constantly waterlogged.

Tolerance to Soils rich in Organic Matter:
This plant prefers soils rich in bases, organic matter, and nutrients. Its presence may indicate fertile and balanced soil, often found in underwoods, hedgerows, meadows, and gardens.

Preference for Neutral Soils:
Viola odorata prefers soils with a neutral pH, ranging from 5.5 to 7.0.

Preference for Shaded to Partially Shaded Environments:
Sweet violet thrives in shaded to partially shaded environments. Its presence can indicate semi-shaded to shaded areas with good vegetative cover in cool climates.

Tolerance to Light Disturbances:
Viola odorata is slightly tolerant to disturbance and can colonize lightly disturbed areas. It spreads through its stolons and will not form extensive mats in environments where the soil has been heavily disturbed.

Indication of Temperate Climatic Conditions:
Sweet violet is common in temperate climatic zones, tolerating moderately cold winters and moderately warm summers. Its presence may indicate a temperate climate with distinct seasons.

VISCUM ALBUM (MISTLETOE)

VIGOROUS HOST TREES
WIDE VARIETY OF SOIL TYPES
PH OF 6.0 TO 8.0
OPEN OR SEMI-OPEN AREAS
VARIED ECOSYSTEM

Preference for Healthy and Vigorous Hosts:

Mistletoe often grows on healthy and vigorous host trees. Its presence may indicate host trees that have sufficient access to water and nutrients. It is commonly found on poplars, apple trees, lime trees, rowans, whitebeams, willows, locust trees, and hawthorns. It is rare on elms and oaks, although the latter is said to possess the most potent medicinal properties. It is very rare on pear trees, cherry trees, hazelnuts, hornbeams, and chestnuts.

Tolerance to a Wide Range of Soil Types:

Viscum album can develop on hosts located in a variety of soil types. The diversity of its hosts and habitats can indicate a wide range of soil conditions.

Preference for Neutral to Slightly Alkaline Soils:

Mistletoe prefers hosts located in soils with a pH ranging from neutral to slightly alkaline, i.e., a pH range of 6.0 to 8.0. Its presence can indicate soils with a stable and often slightly basic pH.

Preference for Sunny to Partially Shaded Environments:

Mistletoe thrives in sunny to partially shaded environments in cool climates. Its presence can indicate open or semi-open areas with good light exposure, necessary for the photosynthesis of mistletoe.

LIST OF PLANTS IN ALPHABETICAL ORDER BY LATIN NAMES

LIST OF PLANTS IN ALPHABETICAL ORDER BY ENGLISH NAMES

DID YOU LIKE THIS BOOK?

Check when more books are coming by following my name *Marie Cachet*. And feel free to leave a comment and a rating!

Made in the USA
Monee, IL
03 November 2026

67377035R00118